Weaving with CLASPED WEFT

Weaving with CLASPED WEFT

30+ Exciting Designs for Towels, Rugs, Runners, and More

TOM KNISELY

STACKPOLE
BOOKS
Essex, Connecticut

STACKPOLE BOOKS
An imprint of The Globe Pequot Publishing Group, Inc.
64 South Main Street
Essex, CT 06426
www.globepequot.com

Photography by Kathleen Eckhaus

We have made every effort to ensure the accuracy and completeness of these instructions. We cannot, however, be responsible for human error, typographical mistakes, or variations in individual work.

British Library Cataloguing in Publication Information available

Library of Congress Cataloging-in-Publication Data available

ISBN 978-0-8117-7392-8 (paper : alk. paper)
ISBN 978-0-8117-7393-5 (electronic)

∞™ The paper used in this publication meets the minimum requirements of American National Standard for Information Sciences—Permanence of Paper for Printed Library Materials, ANSI/NISO Z39.48-1992.

First Edition

CONTENTS

Preface viii

Introduction 1

Understanding the Basics of Clasped Weft 4

A Word about Tools 6

Acknowledgments. 109

Towels

Goose Eye Twill Towel
10

Zanshiori Towels
13

Pacific Northwest Towel
16

Circuitry Towel
19

Emerging Stars Towel
22

Variations on a Theme:
One, Two, and Three
25

Vonnie's Towel
28

Independence Towel
31

Hydrangea Twill Towel
34

Rugs

Three-Colored Towel
37

Blue Blocks Rug
42

Inclines Rug
45

Snow Peaks Rug
48

Kaleidoscope Rug
51

Blue by You Rug
54

Crocus Rug
57

Volcano Rug
60

Wearables

Purple Rain Scarf
64

Naturally Neutral Scarf
67

Cityscape Scarf
70

Bink's Scarf
73

Blueberries and Grapes Shawl
76

Autumn Snuggles Shawl
79

Table Runners

A Quake's a Coming! Table Runners
84

Optical Illusion Table Runner
87

Morning Has Broken Table Runner
90

Puzzle Pieces Table Runner
93

Bits and Bobs

Weaver's Secret Baby Blanket
98

Ocean View Pillow
103

Reykjavik Placemats
106

Preface

Understanding how different weave structures work and then teaching them to others has been the biggest part of my professional career. I have to admit, from the very start, I have been, metaphorically, like a greyhound chasing a mechanical rabbit on a racetrack. I have worked to discover, study, and sample, sample, and sample some more, to fully understand how a particular weave structure works. The more complex and difficult it was, the better I felt about myself for grasping the concept and weaving a project through to the end to say that I did it. And I did this with no regrets. I promise. But then, one day, things changed and took me in another direction. I am so grateful.

I was attending a weaving conference and went to the marketplace to look around. One vender was selling fabrics and clothing from all over the world. There I found the most interesting piece of rag woven fabric. It was an obi that was intended to go around a kimono. It was rag weaving like I had never seen before. I was very familiar with weaving rag rugs, but this was different and brilliant. I purchased the obi, and when I got home I studied this simple woven piece and found areas that were woven as clasped weft. The weft rags were cut very fine to reduce the possible bulk that can happen with clasped weft. I was intrigued, inspired, blown over—and fell in love with this humble piece of rag-woven fabric. I needed to understand more and explore the possibilities with this technique. And so I sampled and wove and wove some more until I understood the possibilities and the restrictions of clasped weft. I am here to tell you that there are very few restrictions and way more possibilities for you to try.

One of the first clasped weft pieces I wove in this journey of exploration was a rug for my last book, *Rags to*

The obi that inspired my journey into clasped weft

Rugs. I planned it for a friend. She loves the warm colors of fall, so for her rug I used rags in brown, rust, and neutral shades and wove in a bold geometric pattern. From there, I began to experiment more and more with clasped weft.

Although I was enamored of clasped weft, I would still ask myself, "Why do you think that this is important, and for heaven's sake, why do you think that anybody would be interested enough to read this book?" Because clasped weft is so very basic and simple to weave, it's often overlooked as a technique and rarely taught even in a beginner's weaving class. Many weavers seem to come across clasped weft in their quest to gain knowledge of textiles and woven structures, but they don't seem to use it very much. The "why" for me is that clasped weft can be woven on all types of looms, including rigid heddle, inkle, table, and floor looms. Whether you have a two-shaft loom or a multishaft Compu-dobby loom, you can add fabulous elements to your woven pieces with some knowledge of clasped weft.

As the saying goes, nothing is really new. I have learned that many cultures around the world have discovered and used the clasped weft technique to create beautiful patterns in their woven textiles. When you are

My friend's clasped weft rug from *Rags to Rugs*

weaving with a loom with limited capabilities, such as a two-shaft loom, you are pushed to the furthest ends of your creative being to make pretty fabric. The use of stripes, both horizontally and vertically, is just one option. Let's add clasped weft to your repertoire.

In this book, I have presented the pieces as projects for you to try and change up to make them yours. I have not included a suggested warp length because I can't possibly know how long or how many items you want to make. You must be the judge of that and do a little work on your own to decide how long the warp should be.

I have written other books on weaving. Some are theme oriented, some are pattern books, and all have project ideas to inspire you. I even wrote two children's books to bring the love of our craft to our littles. I am very proud of my past writing, but this has been an adventure and so much fun to do. I now present to you my study in clasped weft.

Introduction

Hello there. Welcome to this exploration of clasped weft.

I have heard this form of weaving called the interlocking weft technique or "meet and separate," but the name that resonates with me best is clasped weft. In this technique, two or more weft yarns come together from opposite directions. The weft yarns meet, wrap around each other (or clasp), and then return to their original starting point. This is mostly done in an open shed. The point at which the clasping takes place can be shifted to the left or right (or any place where the weaver wants the join to fall).

There is a second way of weaving clasped weft technique that many rug and tapestry weavers use. The wefts start on opposite selvedges and are woven toward each other within the same shed. At the point where the wefts are to meet, the weaver brings the wefts to the surface, up and through the raised warp threads. The weaver then clasps the yarns and returns them to the selvedge edge in a new and opposite shed.

When I have tried to explain clasped weft to other weavers, I have found this analogy quite helpful: Imagine a square dance. Picture partners dancing toward each other, locking arms as if to do-si-do, and then returning to their original position. This is how clasped weft is woven.

Let's take a look at a photo of towels (pages 2–3) woven with the clasped weft technique. Some of you may have seen this style of weaving before and didn't know what it was. The back-and-forth, jagged lines might remind you of a mountain range or even a cityscape. Beautiful patterns can be woven with just two different colored weft yarns woven into a solid color warp.

Over the years, I have woven several scarves and a set of placemats in clasped weft but didn't put too much serious thought into the possibilities that could be achieved. Then, one day, thoughts of new possibilities of clasped weft poured into my head as if a dam had broken. The ziggety zag patterns that I was most familiar with paled in comparison to the new ideas. I thought, *What if I add stripes to my warps and weave to add carefully placed geometric patterns to the design where the joins meet up at the edge of the stripe?* I then considered the possibilities of diagonal movements, and what about curves instead of the sharp points that I usually associate with clasped weft? How about incorporating twill into the design instead of plain weave as I usually do? And what about overshot or other weave structures woven with clasped weft?

I was on a roll over the next few months, trying to weave samples of these inspirations that kept coming into my head. Ideas came flowing from dreams at night, in the shower, while on vacation, and it seemed always at the most inopportune time—like while standing in the street and seeing a truck with fabulous graphics on the side panel or while at a concert, noticing a beautifully patterned dress worn by a concertgoer. I would try to focus and remember what I saw and jot it down as fast as I could so as not to forget this clasped weft opportunity. A pen in my pocket and an old envelope saved many great ideas from taking flight forever.

A trip to the beautiful country of Japan had a great influence on me, too. The Japanese view of aesthetics worked its way into my thought process, and you will see that many of the pieces I wove for this book show these influences. Certain color choices and the use of asymmetrical placement of a design were sometimes difficult for me to get my Western-thinking head around. I never thought too much about the color of koi fish before that trip, but now it's hard to erase the use of orange, gold, white, and black from my head. It's a far cry from the blue and green palette that I am more familiar with.

With each project, I have added some written commentary to accompany it. I share what my influence was for the project and include what I learned from weaving it and what you will want to look out for. In the case of some of the rugs, I will suggest how to prepare the weft materials and make suggestions on the finishing of your piece. For some of the towels, I will give you several ways to weave them. These suggestions come from the results of my own experiments.

In some cases, the drafting notations may be a little different from what you are accustomed to seeing and following. I thought a lot about how I was going to explain this to a reader who was inspired to try one of these projects. I have done my best to make it as clear as possible. You see, you will have two different threads coming to meet, clasp, and then return to the selvedge edge where they originated from. In conventional project drafts, your weft follows a treadling order and the weft thread moves from left to right and then back again. Take your time to think about how this is coming together.

As you look at what I have woven, you will get a sense of how I came upon the woven design. Everything is woven like a stream of consciousness. In most cases, there wasn't a need to measure or keep track of what I had just woven. I just went with the flow until I reached a woven length of what I thought it should be. Other weaving patterns give you a written plan with a materials list, threading draft, sett, and yarn amounts needed. If you follow the pattern, it will look just like the photo that inspired you to weave that project. Clasped weft is a little different because you are going to choose where you place the clasped joins. This process puts you in control of how your piece will look. Ten, twenty, no, a hundred people could follow these suggested drafts, and each one will look a little different. How cool is that?

I hope as you look through these pages you are inspired to weave some of the projects I have included and, of course, put your own spin on them. Try different materials based on your needs. A wool shawl might not be suitable in a warm climate, so change out the yarns for something similar in size and that you will be able to wear. Make it your own. Most important, know that there is no right or wrong when it comes to your clasping choice. Go wild or be as restrained as what you think your piece requires, but always remember to have fun with your weaving.

Happy weaving.

Towels woven with the clasped weft technique

Understanding the Basics of Clasped Weft

Having a better comprehension of how clasped weft works will help you understand my project drafts and instructions. It will also help you later when designing your own projects and deciding how you want to approach the weaving. There are several considerations to keep in mind, and I would like to explain them to you here. That way, you can make wise choices on how you are going to approach your weaving and make it successful. For this book, I have woven numerous samples, in all kinds of different ways, so that I could learn from these samples and then pass that valuable information on to you. Some were successful the first time around; others, not so much. Nevertheless, I learned as much from my mistakes—and what NOT to do next time—as from my successes. Let me share what I have learned from these samples.

In the introduction, I explained that the clasped weft technique is done by having different wefts start at each selvedge edge. They are then woven into an open shed, meet at the weaver's discretion and lock together, and finally return to the selvedge edge where they started. This method will not only place two different weft threads in the same row but also double each strand of weft, making them twice their original thickness. Now let's take into consideration that you are trying to weave a balanced weave fabric. A balanced weave fabric is one in which warp ends per inch (EPI) and wefts picks per inch (PPI) are the same. The weaver's beat also plays a part in the balancing of the warp to weft. If you are using the same threads in both directions, you have a slight problem. The double weft threads in the shed will throw the fabric off balance. You will have fewer weft picks per inch than you do warp ends per inch. This issue can be remedied by weaving with a finer weft thread. A thread half the size of the warp thread is perfect. You will see in some of the pattern drafts that a warp of 8/2 cotton is woven with a weft of 16/2 cotton. That is because I wanted the finished fabric to be balanced. If I had used the same thread for warp and weft, well, then the fabric would be unbalanced, and it would have a heavier hand when you hold it. You might be fine with that result; if so, continue. If you insist on a balanced weave, you will have to purchase more thread of a finer size to make it balanced, which means more expense. "Is there nothing else I can do?" you ask. Why, yes, there is, and I am so glad you asked.

We can easily see that the problem is having two threads sharing the same shed, but what if they returned to their home selvedge edge in a new shed? To do that, the two shuttles would have to start on opposite selvedge edges. They would come together in an open shed and meet at the point where you decide they should clasp together. At this point, you will bring the shuttles to the surface, clasp the two ends, and then change to a new shed. Return the two shuttles into the opening provided by the new shed and toss the shuttles back to their home selvedge. I find a slim, low-profile shuttle is best for this task. A damask-style shuttle with a paper quill, rather than a bobbin, works very well. The side flanges of a bobbin sometimes get caught on the warp threads as you try to come out and then go back into the new shed. Bluster Bay shuttles (Figure 1) and Glimakra shuttles (Figure 2) are two of my favorites for this application. For weft emphasis and rag weaving, this approach is best. The use of a new shed prevents excess buildup of the weft thickness and allows the weft materials to cover the warp better.

If you are planning a warp-faced or warp-emphasis-type fabric, there is no need to use the clasped weft technique. The warp threads are sett closely to make warp patterning the main goal. The clasping of two different wefts would not add any significant part to the pattern.

Now, I know that many of you don't enjoy sampling. I get it, I truly do. "Let someone else do it and then tell me what they found out," you say. I enjoy sampling, and, just to let you know, I have been sampling throughout the writing of this book, and I am going to share and tell you what I have found out. I will share the good and the not so good. Here is an example of just that sort of sampling.

One day, I was weaving a towel, using the usual clasped weft technique that I first described. The double pick ends in the same shed I found to be a bit too heavy for my liking.

Figure 1 Bluster Bay shuttle

Figure 2 Glimakra shuttle

I was treadling for a plain weave fabric. I had threaded the loom with a straight draw twill threading, so it was easy to treadle the odd shafts against the even-numbered shafts to get a simple plain weave fabric. But the fabric wasn't balanced because the warp ends alternated one by one end and the weft had two weft threads in each pick, much as you would find in a basket weave. I took a moment to think about this situation and then treadled for a basket weave so that two warp ends alternate against two adjacent warp ends. That was shafts 1 and 2 against 3 and 4. That day I heard the angels sing, "Ahhhhhhh!" Basket weave was the answer to a true balanced weave fabric when you are using the same size threads for both warp and weft and weaving with a clasped weft.

So here are a few things to think about as you set off on your adventure with clasped weft. Remember, you don't have to set off on your own right away. Let me be your guide through several projects that you might like here in my book. Change the colors, change the threads, make this your project based on your needs. You might like the wool shawl project, but perhaps you live in New Mexico. Please change out that wool for a nice soft cotton or cotton blend yarn of the same weight. It will work just fine for you.

Now say to yourself, "There is no wrong." A towel that is a bit too heavy is now repurposed to a changing pad or a mat for under the dog or cat's bowl. You'll find a use for your wonderful work of art. Have fun!

A Word about Tools

It's important to talk about the tools needed for some of these specialized projects, especially the projects that require the shuttles to come up and out of the open web of the warp. This action could stretch the warp ends and cause uneven tension as your project progresses. I have made a list of the tools I think you are going to need. Here we go!

Looms

Clasped weft can be done on many types of looms. It doesn't have to be woven on a floor loom. It can be woven on a simple frame loom in your lap or a rigid heddle loom if the project doesn't require heavy beating. A scarf or shawl can easily be woven on a rigid heddle loom. Something requiring a heavy beat, such as a set of placemats, a table runner, or, of course, a rug, will need to be woven on a floor loom.

I tried to include a variety of projects to make this book inclusive to many weavers. Many of the projects need only the capability to work plain weave, but some require a four-shaft loom to weave twill and twill-based structures such as overshot.

Shuttles

It's good to have a variety of different types of shuttles: boat shuttles, rug shuttles, ski shuttles, stick shuttles, and so forth. Let's take a moment to review the purpose of each type of shuttle and how they can be helpful in clasped weft weaving.

A boat shuttle is used to carry finer weight thread through the shed as you're weaving. I prefer to use a boat shuttle with a blunt nose so it doesn't accidently pick up warp ends from the bottom layer or top layer of the open shed (Figure 3). A boat shuttle with a pointier nose will sometimes catch a thread in the warp (Figure 4). I have come to love using a slim boat shuttle with a paper quill because they can be easily moved in and out of an open shed while clasping the wefts. The slim shuttle (Figure 5) does little harm to the warp as it is being manipulated during the weaving process. Ski, rug, and stick shuttles are perfect for weaving heavier materials

Figure 3 Shuttle with blunt nose

Figure 4 Shuttle with pointy nose

Figure 5 Slim shuttle

Figure 6 Rug shuttle

such as rug wools and rags. I find using a rug shuttle like the one in Figure 6 works smoothly without any warp catching as you throw it.

Figure 7 Metal and wooden temples

Temples

For many of the projects shown here in the book, I used a temple to help control the selvedges. When clasping the wefts, there is a great deal of pulling against the weft threads, and the selvedges take the brunt of all this clasping and placement of the join. A temple greatly aids in this process, and a temple helps to maintain the fabric at its original width in the reed. I use wooden temples for finer weight fabrics, but for rugs I prefer using a metal temple. Temples come in many sizes, and they are adjustable within their narrowest and widest range.

Thread Stand

When weaving clasped weft, the process can often be woven using just one shuttle. That shuttle carries the weft thread, let's say from the right selvedge edge, toward the opening on the left where it meets up with the opposing weft thread and clasps. The opposing thread can be delivered from a cone sitting on the floor on the left of the loom. To keep the cone from tipping over, a thread stand will do nicely. A large coffee can or other container works well when working with a ball of yarn to prevent it from rolling around the floor.

Figure 8 Thread stand

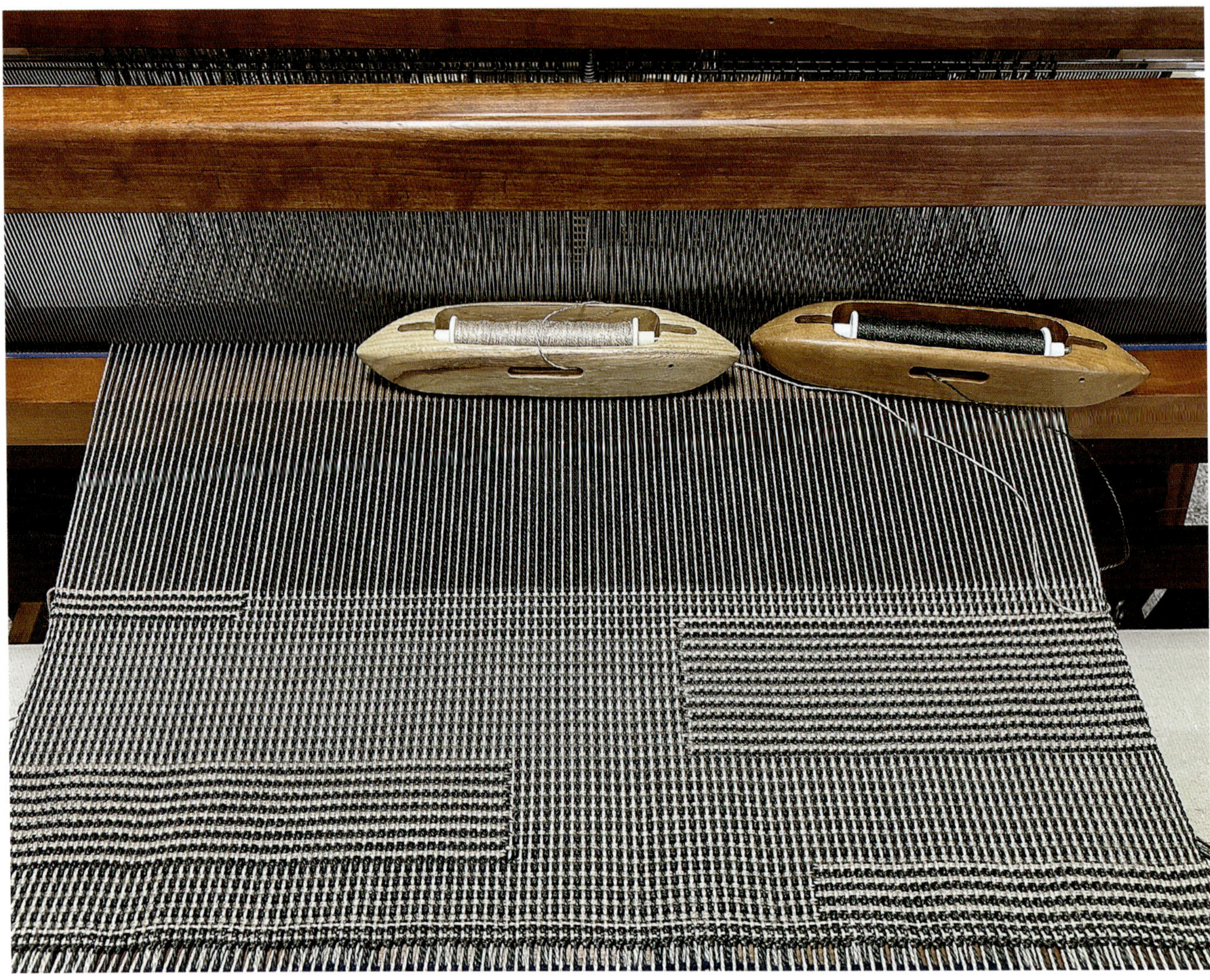

Fringe Twister

A fringe twister is a lifesaver when making a fancy fringe edge. It twists groups of counted threads in one direction and then retwists them in the opposite direction, resembling a spun and plied yarn. Yes, you can do this by hand. For many years, rug weavers used this technique to add beauty and longevity to the fringe ends of their work. Twisting fringe by hand takes a toll on your hands, though, and you will soon feel the cramping. A fringe twister is speedy and efficient and, best of all, a rather inexpensive piece of equipment. Weavers often add fringe for a decorative look to scarves, shawls, blankets, and table linens.

Miscellaneous Tools

These are all the other small tools you will want to have close at hand: scissors, tape measure, pins, needles, bobbins or quills, and a bobbin winder. Use a notebook and phone camera to keep records of what you have done. I learned from one of my students to take a photo on my phone at the beginning of my work. That way, I will have a reminder of what I did in the beginning and can duplicate it on the other end as I am finishing up with my project. The hardest part is remembering to take the photo.

Before you start, gather all the things you will need to work as efficiently as possible, which will make the weaving fun.

Towels

Goose Eye Twill Towel

For this towel, I decided to go with a slightly larger pattern than is appropriate for a piece of fabric such as a dish towel. I felt a small point twill might be a little too delicate for what I was trying to convey. I chose a goose eye twill that has a repeat of 22 warp ends to the pattern. With a warp of 8/2 cotton and a sett of 20 epi, there is slightly more than 1 inch in width to the repeat of the pattern.

Because the clasping of the two weft threads makes for a heavier weight of yarn, I adjusted the treadling to achieve a diamond shape that is not too elongated vertically. The treadling follows a different order than the threading—it is not treadled tromp as writ.

I thought I would try clasping the wefts so that they follow the natural twill line of the pattern rather than going sporadically back and forth, as you often see in clasped weft. This careful placing of the clasps slowed me down considerably, but I think the pattern is really pleasing and worth the time to do it.

I had no particular plan when I started. I wove, moving the line a few inches to the left and, when I felt like it, changing the direction the other way. Back and forth I wove until I reached the length I wanted for the towel. I treated this project as a simple piece of woven fabric with no special consideration for hems. The hems were rolled and stitched down. You might want to do something different for your hems, such as plain weave or a twill with a single thread instead of the doubled threads as I did. Remember, there is no wrong way. Weave what pleases you.

		4				4				4				4				4			
	3				3				3		3				3				3		
2				2				2				2				2				2	
			1				1						1				1				1

				T	T	
		4	4		4	
	3	3		3		
2	2				2	
1			1	1		
				X		Plain
					X	Weave
X						
	X					
		X				
			X			
X						
	X					
		X				
			X			Repeat
X						
			X			
		X				
	X					
X						
			X			
		X				
	X					

Warp

8/2 cotton: Ivory

Total ends: 440

Sett: 20 epi

Width: 22 inches

Weft

8/2 cotton: Black and Burnt Orange

Zanshiori Towels

"Zanshiori" is a Japanese term that translates to "leftover thread weaving." The leftover threads that I use for Zanshiori come from the back of the loom when I have finished a towel warp. They are known as thrums. Thrums are removed from the loom and the ends of the thread are tied together and wound into a ball to be used later as weft. Any knot can be used to tie the threads together. The tails can be trimmed close or left to add a textural element to the cloth. The thrums can be all the same color or a mix of different colors. You can even wind a ball of Zanshiori thrums with different weights of thread and materials. There are no rules to follow. It's all up to you and what you imagine to be good looking.

My wife, Cindy, loves weaving towels. She seems to have several looms going at the same time, and each warp is a different pattern and uses different color combinations. This approach produces lots of thrums of different colors. Our weaving friends will sometimes save their 8/2 cotton thrums for us, too. It doesn't take long to accumulate many bags of thrums. We will sit and tie the lengths together, one at a time, to make a continuous length of thread. It's a good TV-watching or riding-in-the-car activity. We even took bags of thrums on vacation with us and tied them during a lengthy tour bus ride to break up the time. When tying the thrums together, I try never to put two threads of the same color together. Mixing the colors up is the way I usually like to approach the task. But don't spend too much time thinking about it. It should be fun. Because the threads can be of different lengths, the color shifting in the weft is always surprising and serendipitous.

I really like recycling thrums and wondered how I could include this idea in this clasped weft book. It came to me that if I tie only warm colors for one ball of yarn and tie only cool colors for the second ball of yarn, the woven fabric could be quite interesting as they meet and separate in the weft direction.

For this experiment, I wound a warp for two towels in a medium charcoal gray cotton. The warp would provide a neutral background and wouldn't compete with the colors in the weft direction. I also thought that a simple plain weave structure would be best and not look fussy.

The first towel was woven to make a staggering block pattern. The blocks were woven by stacking the clasps on top of each other to make a clean vertical line. To weave a different block, simply shift the vertical line to the left or right depending on how you want your design to develop. Continue this way, weaving the blocks to a desired height. You can shift

Zanshiori Towel 1
Zanshiori Towel 2

them on a diagonal, as you see here, or you can move them back and forth to suit your own taste.

The second towel was woven by randomly shifting the clasped join back and forth horizontally. The color effect is much more subtle. When you look at the towel more closely, you can see that one side of the towel is cooler colored with different values of blues and greens. The other side of the towel has more reds, yellows, and oranges—the warmer colors.

Both towels are equally nice, and the surface textures created by the tiny knots add a lot of interest.

4				4				4			
	3				3				3		
		2				2				2	
			1				1				1

	4	
3		
	2	
1		
X		
	X	4X
	X	Hem
X		5X
X		
	X	Body of
X		Piece
	X	
	X	5X
X		Hem
X		4X
	X	

Warp

8/2 cotton: medium charcoal gray

Total ends: 400

Sett: 20 epi

Width: 20 inches

Weft

Tied thrums in cool colors and warm colors

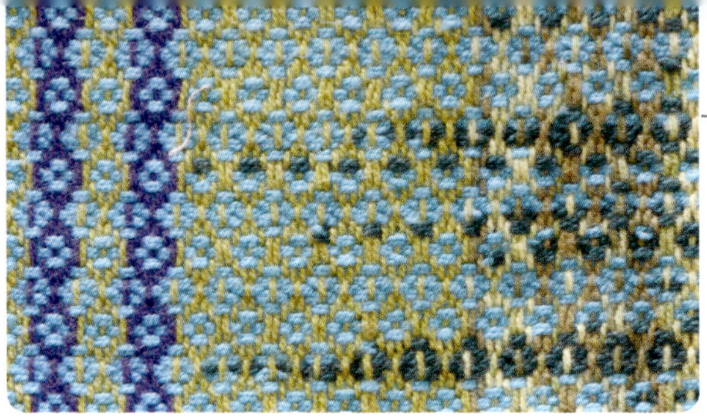

Pacific Northwest Towel

For this experiment, I wanted to weave two towels on the same warp color arrangement. One towel would be woven in plain weave, and the other towel would be woven in a rosepath twill. The Pacific Northwest Towel was woven as twill, and the other towel, called Circuitry (page 19), was plain weave.

For this Pacific Northwest Towel, I chose to use Forest Green and Turk for the weft. The clasps moved back and forth sporadically, and as they did, I could see what looked like blue wildflowers on one side of the towel and moss and cedar branches on the opposite side. The strips of Cayenne running through the green add a great deal of interest and balance to the design in my mind. I hope you like it as much as I do.

You could also stack the clasps, using the warp-way stripes as a guide, and create a totally different look that would be bold and geometric. The possibilities are yours to explore and make your own.

Now take a look at the Circuitry Towel. You can see how they are related and share a common warp but are very different in their appearance.

		4		4			
	3				3		
2						2	
			1				1

46X

Warp Color Order

	Ends											
	104						48		8		40	Limette
	24		8		8							Cayenne
	24							8		8		Royal
	216	104		8		96						Variegated Green
Total ends	368											
				2X					2X			

				T	T
		4	4		4
	3	3		3	
2	2				2
1			1	1	
X					
	X				
		X			
			X		
X					
			X		
		X			
	X				

Warp

8/2 cotton: Limette, Cayenne, Royal, Variegated Green

Total ends: 368

Sett: 20 epi

Width: 18.4 inches

Weft

8/2 cotton: Forest Green and Turk

Circuitry Towel

This towel shares a warp with the Pacific Northwest Towel. Pacific Northwest was woven as a twill, whereas this towel was woven as plain weave. I did this on purpose to see the differences you could achieve on a rather busily patterned warp. The warp is made up of two different green threads—one bright, solid green and the other a variegated green. There are also stripes of bright blue and warm, Cayenne red.

I set the variegated cone of green on the floor to the left side of the loom, and on the right selvedge side I decided to weave with two shuttles. One shuttle carried the Limette green, and the other shuttle carried a bobbin of Royal blue. Through most of the body of the towel, the two colors on the right side alternated with four picks of green and then four picks of blue. These shuttles wove and intersected and clasped with the variegated thread sitting on the floor to the left of the loom. When weaving with the blue weft, I would shift the clasped joins so that they would fall at the edge of a warp stripe. I shifted the joins so that the horizontal stripes would move on an angle and also create a checked area. The four picks of green would intersect with the green variegated thread unless they sometimes intersected with the cone of Cayenne that I introduced on the left side.

This piece was fun to weave because I didn't start out with a set plan. I let it evolve as I went. It was freeing and exciting to weave this way.

		4		4			
	3				3		
2						2	
			1				1

46X

Warp Color Order

	Ends											
	104						48		8		40	Limette
	24		8		8							Cayenne
	24							8		8		Royal
	216	104		8		96						Variegated Green
Total ends	**368**											
				2X	2X				2X	2X		

				T	T
		4	4		4
	3	3		3	
2	2				2
1			1	1	
				X	
					X
				X	
					X

Warp

8/2 cotton: Limette, Cayenne, Royal, Variegated Green

Total ends: 368

Sett: 20 epi

Width: 18.4 inches

Weft

8/2 cotton: Limette, Variegated Green, Cayenne

Emerging Stars Towel

There is a saying that goes, "Crumbs are the thrums of bread, and thrums are the crumbs of thread." I think of this every time I finish a project warp and clean up the waste threads left behind on the loom. For many weavers, this process means a quick removal from the apron rod and then into the trash they go, never to be seen or used again.

For this piece (another Zanshiori weaving), I used thrums from a previous towel warp of 8/2 cotton. The warp had even width strips of Black, White, Brown, Ivory, and Gray threads in the warp. This arrangement gave me the same number of ends of each color when I got started. Each thrum end was around 24 inches long. I tied them in that order and wound them into a ball about the size of a walnut. This became the weft contribution on the right-hand side of this clasped weft towel. I decided to use Ivory for the weft on the left side of the loom. It's the same Ivory as the warp. My initial thought was to weave, clasping the thrums, moving them back and forth to make a jagged edge with simple color changes happening as the knotted ends appeared. Soon after weaving a few inches, I noticed a very disappointing imbalance to the towel's design. The left side of the towel had lots of color changes and textures happening, and the right side of the towel was a wasteland of vanilla-colored plain weave fabric. I thought about how I could perk this side of the towel up, and it came to me to add a textural element by using Brooks bouquet randomly throughout this area. I liked this approach so much better. It elevates the design and makes it more interesting without a lot of extra work.

I have included a few pictures of the progress as I wove. For a good illustration of how to do Brooks bouquet, refer to *The Weaver's Idea Book* by Jane Patrick.

A ball of thrums

4				4				4			
	3				3				3		
		2				2				2	
			1				1				1

	4	
3		
	2	
1		
X		Plain Weave
	X	
X		
	X	

Warp

8/2 cotton: Ivory

Total ends: 440

Sett: 20 epi

Width: 22 inches

Weft

8/2 cotton: Ivory

Thrums in Black, White, Brown, Ivory, and Gray

Weaving Brooks bouquet

Variations on a Theme

4				4				4			
	3				3				3		
		2				2				2	
			1				1				1

Warp Color Order

	Ends								
	200			20		4		60	**Bleu Pale**
	180	60	20		40				**Royal**
	60						4		**Navy**
Total Ends	440								
			4X			15X			

				T	T	
		4	4		4	
	3	3		3		
2	2				2	
1			1	1		
				X		Plain Weave
					X	
				X		
					X	
X						Twill
	X					
		X				
			X			
X						Basket Weave
		X				
X						
		X				

Warp

8/2 cotton: Bleu Pale, Royal, and Navy

Total Ends: 440

Sett: 20 epi

Width: 22 inches

Variation One

I wanted to experiment to see what I could weave on a single warp—a warp that had several different striped areas that varied in color and width. Take a look at the draft to become familiar with my plan and thoughts. I didn't put too much thought into balancing the pattern. I really wanted to make the warp asymmetrical and then weave it with no preconceived plan as to what I would do next. I just wanted to play and weave by the seat of my pants. I started out with the idea that I would weave several different towels. This is Variation One. I started out by placing the Bleu Pale cone on the floor to the left side of the loom and winding a bobbin of Royal and putting it into the shuttle that would weave from the right side of the loom. The Royal shuttle traveled across the warp in a plain weave shed and clasped with the Bleu Pale thread. I then threw the shuttle back to the right side using the same shed, pulling on the Royal thread to place the join somewhere in the wider striped area, and beat it down into place. I changed to the opposite plain weave shed and repeated the process. This time I placed the clasped join in a different part of the wide striped area of the towel. I continued this way throughout the weaving of the towel.

I like the way the towel wove with a narrow-striped pattern on the left side of the towel and an Ikat-looking fabric in the wider striped area on the right side. It could also be described as an inlay, but it's not. The left- and right-side borders of the towel weave as Bleu Pale and Royal.

Note that these projects combine a twill threading with plain weave and basket weave. The treadling chart includes twill for an extra option.

Variation Two

This second towel became a sampler of sorts. First of all, it is a lesson in learning how to carefully place and stack the clasped joins on top of one another to make a vertical line that's clean and crisp. I first started out weaving with the cone of Bleu Pale thread on the left-hand side of the loom and the Royal in a shuttle on the right-hand side. I started out the same way as Variation One, weaving plain weave, but this time I drew the Bleu Pale thread right up to the left edge of the center blue stripe and carefully placed the clasped join right on the line. I continued this way for a few inches, which gave me Bleu Pale and Navy vertical stripes on the left side and Bleu Pale and Royal vertical stripes on the right side of the towel.

I then added a Navy cone to the left side of the loom. For the next several inches, I alternated weaving four picks of Navy with four picks of Bleu Pale while still placing the clasped joins right on the line of the center blue stripe. I carried the Bleu Pale and Navy up the left-hand side of the towel. It wove as a gingham check on the left and vertical blue stripes on the right-hand side of the towel. After several inches of this configuration, I switched to weaving an inch of Bleu Pale with an inch of Navy. This time, I began a new color and ended a color by tucking the ends into the shed. On the right side of the loom, I used two shuttles alternating one-inch stripes of Bleu Pale and Royal to a bold check.

A little later, I went back to using just Bleu Pale on the left side and Royal on the right. This time I shifted the clasp after every four picks to a different vertical Navy warp stripe. I wove from stripe to stripe, and a diagonal stripe began to appear.

When I got to the center blue warp stripe, I shifted my color order and wove with Navy on the left and Bleu Pale on the right. When my angled design reached the center blue stripe again, I shifted my color order once again and wove with Bleu Pale on the left and Navy on the right, shifting the join of clasped weft every four picks.

Variation Three

This towel was also woven on the same warp as Variations One and Two. The difference is that this time I wove the towel in basket weave. It occurred to me that if the weft threads are doubled in the process of clasping two threads, then, if I treadled a simple basket weave, the towel would be balanced in both the warp-way direction and the weft-way direction.

I placed the Bleu Pale cone on the left side of the loom and wound Royal cotton on bobbins for the shuttle on the right side of the loom. The pattern happens when you weave and place the clasped joins on the left side of the vertical stripes in the warp. I first placed the clasps on the left of the center stripe and stacked the joins for several inches. I then moved the clasps to the next vertical Royal stripe, just to the right of the center stripe, and wove for 1 inch. Next, I moved the join to the next vertical stripe and again placed the clasped join right on the line of the left side of the stripe. I continued weaving this way until I reached the large Royal stripe on the right border. The developing pattern reminded me of stairsteps. From here I reversed the pattern sequence until I reached the center stripe.

I continued this way of shifting the clasped joins until I thought my towel was long enough. To make a stable edge to finish, I wove the beginning hem and ending hem in plain weave for an inch and a half so that there was plenty of fabric to roll a hem and stitch it securely.

Vonnie's Towel

Our friend Vonnie is a straight shootin,' no-nonsense sort of person. Vonnie knows what she likes and dislikes and will tell you without any hesitation. That is what I love about this unique lady—there is never a question about how she feels. Vonnie is not world famous or a celebrity, but, like many well-known individuals, she has her own brand. George Burns was well known for his cigars and Iris Apfel for her large designer glasses. Vonnie is known in weaving circles for her love of the color spring green. Vonnie loves spring green and uses it whenever she can. It has become her signature color. We, her friends and admirers, have come to call it simply "Vonnie Green." She even drives a 1979 Volkswagen camper bus in Vonnie Green. When I asked her whether she would mind weaving a contribution to this book, I was thrilled when she said yes. I knew that Vonnie would come through with a beautifully woven piece, and it would have spring or Vonnie Green in it somewhere. I gave her free range to do what she wanted as long as it contained some element of clasped weft. Here is what she has lovingly presented for my book.

Vonnie used an eight-shaft twill threading for this towel. The pattern weaves a checkerboard design with alternating blocks of twill and plain weave. Much like weaving overshot or other supplementary weft structures, this pattern calls for a tabby thread to be woven between the picks of the pattern. It is this tabby section that inspired Vonnie to do the clasping.

The warp is 8/2 Natural colored cotton. The pattern weft is 5/2 cotton in, of course, Vonnie Green (formally known as Pistachio). The tabby threads are 16/2 cotton in Charcoal and Orange. This finer 16/2 cotton is the equivalent to 8/2 when it is clasped and doubled back on itself. The towel, like Vonnie, is a real showstopper.

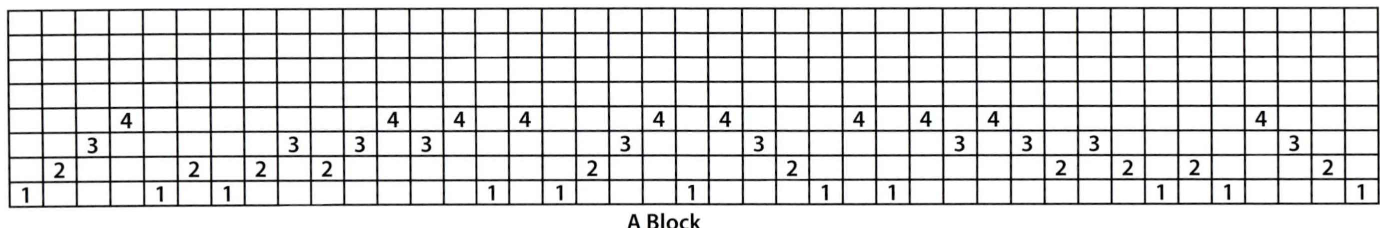

A Block
41 Ends

8				8		8								8		8				8				8		8								8		8				8
	7				7		7		7								7						7								7		7		7				7	
		6						6		6		6						6				6						6		6		6						6		
			5								5		5		5				5		5				5		5		5								5			

B Block
41 Ends

								T	T	
8	8		8			8			8	
7	7	7					7	7		
	6	6	6	6					6	
5		5	5		5			5		
	4			4		4	4		4	
3					3	3	3	3		
			2	2	2	2			2	
		1		1	1		1	1		
								X		Plain
									X	Weave
X										
	X									
		X								
			X							
X										
	X									
		X								
			X							
		X								
	X									
X										
			X							
		X								
	X									
X										
				X						
					X					
						X				
							X			
				X						
					X					
						X				
							X			
						X				
					X					
				X						
							X			
						X				
					X					
				X						

Threading note: *Thread A and B blocks 6 times. Then thread another A block to balance.*

Warp

8/2 cotton: Natural

Total ends: 533

Sett: 24 epi

Width: 22.25 inches

Weft

Pattern weft: 5/2 cotton in Pistachio

Tabby weft: 16/2 cotton in Orange and Charcoal

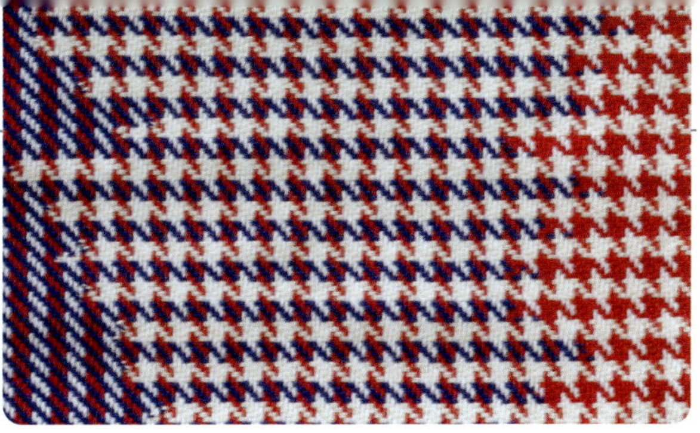

Independence Towel

Don't you find it funny how some things come to you while you are doing something totally unrelated? That was the case with this little hand towel. My wife and I wanted to weave some new napkins for the Christmas season. Cindy put me to the task of designing something simple in red and white. I came up with a striped warp with alternating Red and White cotton and planned to weave the colors to square in a simple twill: four Red alternated with four White threads in the warp and then woven with four Red and four White weft rows to make a pretty checked fabric.

When the napkins were finished, there was still a little warp remaining, and I figured there might be enough to weave a guest towel for our powder room. I thought, since I was working on this book about clasped weft, why not use this opportunity to use up the warp for a simple project that looks much more complex than it really is? The end result is one of my favorite pieces for this book. It is simple, straightforward, and, best of all, easy to weave.

The warp was wound with 8/2 cotton. For the napkins, I used the same 8/2 cotton as the warp in Red and White. Because this towel was going to be woven using the clasped weft technique, I decided to use 16/2 cotton for the weft threads in the colors Red, White, and Royal. Clasping the weft threads in the shed doubles the thickness of weft rows. By reducing the weight of the weft thread by half, the clasped weft is now the same weight as the 8/2 cotton.

The Royal thread stayed on the left side of the loom throughout the entire weaving of the towel. On the right side of the loom, I wove alternating four picks of Red with four picks of White and treadled a simple straight draw twill. You can see that when the Red clasped with the Blue, the clasp was drawn closer to the right side of the fabric. When the White weft was clasped with the Blue, the clasped join was moved closer to the left side of the towel. This created a new section of color study in the middle of the towel.

On the left side of the towel, there are vertical stripes of Red and White with a Royal weft. The right side of the towel has the same Red and White twill check as our Christmas napkins, and in the middle of this sweet towel, there is a whole new pattern created by having a warp of Red and White stripes and a weft arrangement of Blue and White stripes. I'd say, "A whole lot of BANG for your buck." Like Fourth of July fireworks.

We now have a guest towel to put out on our national holiday celebrations. And the best thing is, it was woven on the remaining warp of a different project. Raise a glass to frugality!

		R				W				R				R	
			R				W				R				R
R				R				W				R			
	R				R				W				R		
						50X									

				T	T	
		4	4		4	
	3	3		3		
2	2				2	
1			1	1		
				X		Plain
					X	Weave
X						
	X					Twill
		X				
			X			

Warp

8/2 cotton: Red (R), 208 ends; and White (W), 200 ends

Total ends: 408

Sett: 20 epi

Width: 20.5 inches

Weft

16/2 cotton: Red, White, and Royal

Hydrangea Twill Towel

This towel's name suggests something specific, maybe even a little fussy. The warp is really a blank slate that will accept numerous colors, and it allows for a lot of different treadling possibilities.

I chose a soft, light blue color for the warp. The threading is a rosepath variation with the twill's turning points happening every inch. It's bold and makes a statement.

The weft colors also make a bold statement. The weft colors are Teal and Lime Green. The towel was treadled to a simple twill, and the clasped joins were shifted back and forth, hyper-randomly, to give the appearance of movement.

The towel was finished with a simple rolled hem, washed on a regular cycle in warm water, and machine dried.

Make sure that you put on a nice, long warp. You will want to try different treadling possibilities and different colored wefts and weave lots of towels on this warp.

		4				4				4		4				4				4			
	3				3				3				3				3				3		
2				2				2						2				2				2	
			1				1				1				1				1				1

				T	T	
		4	4		4	
	3	3		3		
2	2				2	
1			1	1		
				X		Plain Weave
					X	
				X		
					X	
X						
	X					
		X				
			X			

Warp

10/2 cotton: Periwinkle

Total ends: 480

Sett: 24 epi

Width: 20 inches

Weft

10/2 cotton: Teal and Lime Green

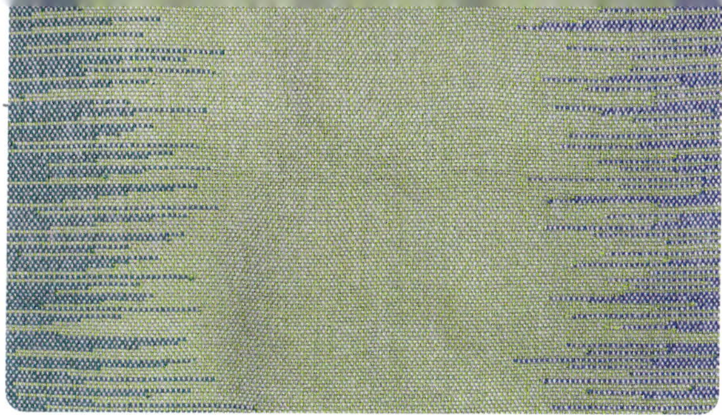

Three-Colored Towel

This towel should look familiar. The colors are the same as the Hydrangea Towel (page 34), and it is woven on the same warp. I love it when I can get several towels from the same warp. There is another color added to this exercise, making it a weave with three colors. The treadling is a simple plain weave. Working with three colors adds a little challenge to the weaving. Let me tell you how to overcome the challenge.

First off, though, a warning: This should not be your first attempt at clasped weft. Start out with a two-color clasped weft project to get yourself familiar with the whole process of weaving with an active shuttle and a single cone of thread sitting on the floor providing your second color.

I started out thinking about the hem. For this, I used the color that would be the center color in my towel: Lime Green, woven singly and unclasped. When it was time to introduce the two new colors, I chose two different blues: Teal and Royal. I placed a cone of each color on opposite sides of the loom—Teal on the left and Royal on the right. The Lime Green color was wound onto a bobbin and placed into the shuttle.

The change from one shed to another shed

Clasped join in the shed

Starting on the right side, I opened the 1–3 tabby shed and threw the shuttle across to the left side and clasped the end of the Teal thread. I then threw the shuttle back to the right side, dragging the Teal thread to where I wanted the join to fall in the shed. I beat the threads down gently to place them. Next, I reopened the same 1–3 tabby shed and clasped the Royal thread on the right, and then I pushed the shuttle gently back into the shed, dragging the Royal thread into the shed. **At the spot where I wanted the clasped join to fall, I brought the shuttle up and out of the open web and then closed the shed and beat. It's at this point that you need to change the shed to the 2–4 shed and reenter the shuttle at the same spot where it came out. Enter the shuttle into the new open shed and throw the shuttle to the left and clasp the Teal thread, and, like before, throw the shuttle back to the right-hand side of the loom. Decide where you want the clasped join to be and gently beat it down into position. Open the 2–4 shed again and now clasp the Royal thread and return the shuttle to the open shed and bring the shuttle up and out of the open web where you want the join to appear.** Beat and repeat the process for the length of the towel.

The towel was finished with a rolled hem, washed on a warm cycle, and machine dried.

Clasping the two weft threads outside the shed

Adjusting the chosen position of the join

Shuttle exiting to top layer of an open shed

The clasped join in place and the green thread brought to the surface

		4				4				4		4				4				4			
	3				3				3				3				3				3		
2				2				2						2				2				2	
			1				1				1				1				1				1

	4
3	
	2
1	
X	
	X
X	
	X

Warp

10/2 cotton: Periwinkle

Total ends: 480

Sett: 24 epi

Width: 20 inches

Weft

10/2 cotton: Teal, Lime Green, and Royal

The next shed with the shuttle reinserted into the new shed at the spot where the join was made

Cones on the floor to the left and right of the loom

Rugs

Blue Blocks Rug

When I was in the planning stages for this rug, I decided to make the warp long enough for two rugs. One would have a bold geometric pattern made up of blocks of blue and a neutral color, and I would weave the other with bold diagonal lines. Clasped weft lends itself nicely to both ideas. To help with this large-scale design, I made the warp with five wide stripes of alternating dark and light-colored areas.

Take a moment to look at the threading draft. The dark stripes are sleyed with two Navy and two Black threads alternating throughout that stripe. That stripe begins and ends with Navy to balance the stripe. I then have a wide stripe of the color Tan. The five stripes are approximately 5 inches wide. The dark stripes are slightly wider because of balancing the colors.

To prepare the rags, I tore them into 1½-inch-wide strips. I reserved several of the dark-colored strips for the beginning and ending portions of the rug. These rag strips would be woven just as they are for the first six picks and the last six picks of the rug. These dark-colored, 1½-inch rags are woven all the way across the warp as you would naturally when weaving a rag rug. The remaining dark-colored strips and all the light-colored strips are then torn again in half to make them approximately ¾ of an inch wide. If you carefully tear the rag strips to within a half inch of the end, you can then trim the corners of the end, and you now have a rag strip that is twice as long as the one you started with. You are tearing these strips to ¾ inch so that when you make the clasp and return the rags back to their selvedges, the folded rag strips are now weaving as a 1½-inch-wide strip in that shed.

To help make the vertical edge of the blocks straight, I used the natural break in the colored stripe as my guide. I began with a blue rag strip on the right and clasped it with the light-colored rag strip coming in from the left side of the warp. I wove this way with no plan as to how high the blocks were to be. I kept it that way throughout the weaving and was spontaneous as to where I wanted the blocks to change. When I was satisfied with the height of the block, I cut the rags' ends to a taper and tucked the ends back into the previous shed that they were woven in. I then changed sides, and the light-colored rags were now weaving from the right and the dark-colored rags were on the left. This is also when I made the decision to relocate the clasp to another vertical break in the warp's color stripe. I wove this way until my rug was as long as I wanted it to be or until I nearly ran out of rag strips.

Remember the wider, 1½-inch torn rag strips in the darker color? I finished the rug with six picks of these wider strips. I did use a hem for the beginning and ending of the rug. I have come to love a rolled hem on a rag rug. Washing and maintaining the rug is so much easier without a fringed edge.

4				4				4			
	3				3				3		
		2				2				2	
			1				1				1

Warp Color Order

	Ends												
	100	2		2		2		2		2		2	Navy
	90		2				2				2		Black
	120				60				60				Tan
Total Ends	310												
			15X				15X				15X		

	4	
3		
	2	
1		
X		4X
	X	
	X	5X Hem
X		
X		6X
	X	
X		3X 6 Picks of 1.5 in. Strips
	X	
X		Body of Rug Woven with 0.75 in. Strips
	X	
X		
	X	
X		
	X	
X		3X 6 Picks of 1.5 in. Strips
	X	
X		6X
	X	
	X	5X Hem
X		
X		4X
	X	

Warp

8/4 cotton carpet warp: Navy, 96 + 4 = 100 ends (double first two and last two ends); Black, 90 ends; and Tan, 120 ends

Total ends: 310

Sett: 12 epi

Width: 25.5 inches

Weft

45-inch-wide batik fabric, 6 yards each in light and dark prints

8/4 cotton: Black and Navy wound together for the hems (a double bobbin shuttle works well for this process)

Inclines Rug

The draft for this rug is the same threading draft as for the Blue Blocks Rug (page 42). I mentioned in the description for Blue Blocks that I had two rugs planned when I made the warp. Since I was warping for one rug, why not make it long enough for two rugs? Everything else regarding the hems and beginning process for this rug is like the Blue Blocks Rug except for how I wove the diagonal inclines and declines for this rug. Let me tell you how I approached the weaving for the Inclines Rug.

I loved the idea of weaving diagonal lines, but the idea of using a protractor to achieve a correct angle is far too fussy for this old weaver. My personality is much too *wabi-sabi* for me to care that the angle is a perfect 45 degrees. I just wanted the colors to ascend and descend to make points. I took a deep breath and just started weaving. Now I knew from past experiences with tapestry weaving that I would need to move my color joins slightly to the left or right with each consecutive weft pick to get the angular slope that I needed. Not like I did when weaving Blue Blocks. There, I stacked the clasped joins directly on top of one another and along the color transition in the warp.

Because the rag strips are bulky and not like using yarn, I had to play around and sample a few picks to see just how much I would have to shift to the left or right to get the angle I wanted. First, I found that if I moved the join just two warp threads over to the right with each pick, the angle was going to be very steep. I tried moving over four warp threads with each pick of the rags, and this worked out so much better. Voilà! I had found the combination from this sampling, and so I stuck with shifting the join every four warp threads with each successive row of the weaving of the rug.

I started the pattern by clasping the first join a few inches in from the edge of the rug. I wove, moving the pattern angle first to the right and then shifting the angle to the left, making what looked like a large, blue pointed puzzle piece. I then decided to switch sides and started the colored rags on the opposite side of the rug so that the blue point would now appear on the other side. I continued this way for a while and then switched back again to the original side to finish the rug and balance out the design.

I found the whole process fun to weave without the pressure to achieve a certain angle in the design. Just let it happen naturally and take shape according to your material choices and your beat. Remember that there is no wrong way to do this.

4				4				4			
	3				3				3		
		2				2				2	
			1				1				1

Warp Color Order

	Ends												
	96	2		2		2		2		2		2	Navy
	90		2				2				2		Black
	120				60				60				Tan
Total Ends	306												
			15X				15X				15X		

	4	
3		
	2	
1		
X		4X
	X	
	X	5X Hem
X		
X		6X
	X	
X		3X 6 Picks of 1.5 in. Strips
	X	
X		Body of Rug Woven with 0.75 in. Strips
	X	
X		
	X	
X		
	X	
X		3X 6 Picks of 1.5 in. Strips
	X	
X		6X
	X	
	X	5X Hem
X		
X		4X
	X	

Warp

8/4 cotton carpet warp: Navy, 96 + 4 = 100 ends (double first two and last two ends); Black, 90 ends; and Tan, 120 ends

Total ends: 310

Sett: 12 epi

Width: 25.5 inches

Weft

45-inch-wide batik fabric, 6 yards each of light and dark prints

8/4 cotton: Black and Navy wound together for the hems (a double bobbin shuttle works well for this process)

Fabrics used in weft

Snow Peaks Rug

You never know when you are going to be inspired by something you see. As I stepped out of the hotel elevator, I took notice of the carpeting in the hallway leading to our room. The carpet had a pattern of sharp zigzag lines running lengthwise down the hall. It was light colored on one side and dark on the other side. The intriguing thing about the pattern areas was that they were not solid colored but heathered. I thought, "How simple this is and how strikingly beautiful." I knew I could weave this design in a smaller version.

When I got home, I started looking for rug wools that would work for this new rug. I like the rug wools from Maurice Brassard. They are well-spun, 3-ply rug yarns. To get the tweedy, heather look that I wanted, I would have to strand several different yarns together on the shuttle and treat them as one yarn. These stranded yarns were going to make a very bulky weft yarn. Because the rug's design lends itself to being weft-faced, I would need a rather coarse sett to get a weft-faced or weft-emphasized rug. I choose a sett of 4 working ends per inch. This open sett allows the bulky weft yarns to pack down tightly and cover the warp. For the warp, I used a natural cotton seine twine in size 12/6. I doubled the ends in the dent and also in the heddle eye.

For the light-colored areas in the rug, I chose to wind a natural white, a sable, and a light tweedy gray color together on a shuttle. For the dark areas, I wound a black yarn, a charcoal yarn, and the same color gray that I used for the light-colored area. Using the same gray yarn for both pattern areas helped to soften the look of these sharp, jagged points coming together.

4				4				4			
	3				3				3		
		2				2				2	
			1				1				1

	4
3	
	2
1	
X	
	X
X	
	X

Warp

12/6 cotton: Natural

Total ends: 200

Sett: 8 epi (4 working epi); double the ends in the dent and in the heddle eye

8

Width: 25 inches

Weft

Maurice Brassard 3-ply Rug Wool: Black, Dark Gray, Light Gray, White, and Sable

The hotel carpet that inspired this rug

Kaleidoscope Rug

My friend Betsy travels to India quite often. I asked her to bring me back some cotton saris sometime, but only if she had room in her suitcases. Cotton saris make a nice, long length of lightweight fabric and are often decorated with lovely metallic designs. I planned to use them to make fabric strips for rag rugs. Betsy surprised me one day with bags of new and slightly used cotton saris, and the colors were bold and vibrant. They were just what I was hoping for.

Unsure of the colorfastness, I washed all the saris in hot water with color catchers and dried them in a hot dryer. I was surprised at how little of the color came out. The colors retained their brightness, although I did lose some width and length in the fabric. It didn't really matter to me because I was planning on tearing the fabric into long strips and weaving them into a rug. If the fabric from one sari ran out, I simply used fabric from a new sari. With some saris, I had plenty of strips left over for another rug. I used fabric from five different saris and from time to time included some of the metallic strips to give the piece a surprising bling.

The warp was made from a natural-colored seine twine known as Fishgarn. I sett the warp at four working ends per inch to allow the rags to beat down and mostly cover the warp. There were areas that didn't completely cover, but that all adds to the look of the rug. In these areas, the rags were torn a little wider, which made them thicker and harder to pack down.

The clasping technique for this rug is a little different from others in this book. I opened the shed and wove rags into the shed with them coming from opposite sides as usual. Where I wanted the clasped join to take place, I brought the shuttles to the surface by pushing them up and through the upper layer of the web, and then I beat the rags down against the fell. Next, I opened the opposite shed, wrapped the two rags around each other to clasp them, and then reinserted them into the opening that they previously came out of. I subsequently pushed the rag shuttles back to their original starting point. I beat these rags down into place and then started the whole process over again.

I wanted a rug with bold geometric blocks—something that resembled a crazy quilt. There wasn't a plan when I started. I let the pattern develop as I wove. The only pre-thought to this piece was to weave bold colors against somber ones and to weave cool colors against warmer tone rags. I shifted the clasps from side to side and wove the blocks at different heights to add interest. I like the way it evolved.

I finished the rug with a Damascus edge and then twisted the fringe to give it lasting strength.

4				4				4			
	3				3				3		
		2				2				2	
			1				1				1

	4
3	
	2
1	
X	
	X
X	
	X

Warp

12/6 Fishgarn (seine twine): Natural

Total ends: 200

Sett: 8 epi (4 working epi); double the ends in the dent and heddle eye; use an 8-dent reed

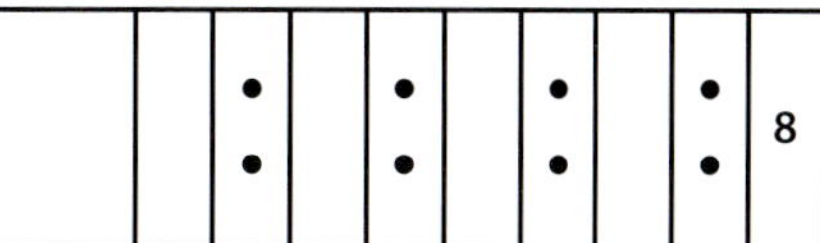

Width: 25 inches

Weft

Cotton sari fabric cut to 1-inch-wide strips

Blue by You Rug

This was a fun way to use up several pieces of blue patterned batik fabric. The odd pieces of fabric were all different lengths and varied in their pattern colors. What they did have in common was the color blue. Some were light blue, some were blue and turquoise, and one had pink and green dots within a predominately blue background. My thinking was this: I will use them all, but I will never allow the same fabric prints to clasp together from opposite ends. That would be silly because it would read as if I had the same fabric weaving all in one horizontal line. Some of the rag strips were long, so I clasped that color with two or three shorter-length rag strips of different colors coming in from the opposite side of the loom. I stacked the clasped areas to give straight vertical lines and produced a design that was bold and geometric.

I started by tearing the batik fabrics in 1½-inch-wide strips. I would normally tear these fabrics into narrower strips, but I wanted this rug to be heavy and thick. It worked out to be exactly what I was looking for, but it took a little rethinking as far as my treadling sequence was concerned.

I warped the loom with Dark Brown 12/6 cotton from Berga Yarns. It is sometimes called Fishgarn. It has a tight twist and is very strong and smooth. The warp was sleyed for 8 epi, but the treadling sequence makes these threads work as 4 working epi. This approach makes for a courser and wider sett, and the clasped joins can fall down in between the warp ends.

I planned for a rolled hem. The hem was woven in plain weave. I used a blue cotton yarn called Sugar and Cream. I wove the hem for about 2 inches, which was enough to make a sturdy hem and the color blended in nicely with the blue batik fabrics in the rug.

SPALDING-OLYMPIC

4				4				4			
	3				3				3		
		2				2				2	
			1				1				1

				T	T	
		4	4		4	
	3	3		3		
2	2				2	
1			1	1		
				X		Hem
					X	
				X		
					X	
X						Body of Rug
		X				
X						
		X				
				X		Hem
					X	
				X		
					X	

Warp

Berga 12/6 cotton: Dark Brown

Total ends: 200

Sett: 8 epi

Width: 25 inches

Weft

Several blue batik fabrics cut to 1½-inch-wide strips

Fabrics used in weft

Crocus Rug

For this rug, I wanted to use yarn that was available in most yarn or craft stores. I also wanted yarn that was cotton, as well as inexpensive, and could be easily washed without a whole lot of fuss. It also needed to come in lots of colors so that it would appeal to a lot of weavers. My wife is a weaver and a knitter and likes the yarns Sugar'n Cream and Peaches & Creme. These cotton yarns have a soft twist and are the size of what I would call a worsted weight yarn. Cindy let me sample with yarns from her stash, and I have to say it was just what I was looking for. The yarns beat down into a very tightly woven rug, and the process of clasping these yarns was very easy.

I decided to use a wide sett, much like you would use for a tapestry, so that the weft yarns would beat down tightly and cover the warp. Even with clasping and doubling the weft yarns, the Sugar'n Cream beat down and completely covered the warp threads.

My color choices were based on the colors that are predominant in our living room. I used solid colors of teal, blue, green, and lavender. I also used a variegated yarn with greens and blues. It was an experiment for me. I wanted to see how the variegated thread would change the pattern depending on the length of the weft as it clasped with a solid-colored area. Longer lengths of the variegated weft seemed to look like striations of color, and the shorter lengths sometimes stacked the colors so that it gave vertical stripes. Sometimes the pattern was a pleasant surprise that's hard to describe, but I liked it and kept going on. I have to tell you that I didn't attempt to match the colors of the variegated yarn when a bobbin ran out and I joined a new bobbin. If the color of the ending bobbin was white and the beginning color of the new bobbin was blue, I simply joined them at that point. I do taper the ending and beginning yarns. Simply tear two strands away from the four-ply yarn. Doing so reduces the bulk as they overlap.

I also thought it might be interesting to clasp the colors so that they resembled narrow finger joints. When I found the spot where I wanted the clasped join to fall, I stayed in that spot for a ten-pick repeat. I then moved to a new spot, either to the left or to the right of the last join, and did the same ten-pick repeat. The solid colors changed at my discretion. This approach kept the design more geometric and more like a stream of consciousness than an EKG heartbeat graph.

For the finished edge, I decided not to have a fringe. There were just too few warp ends to make a nice-looking fringe. I started with a half-Damascus edge, which produces an edge of half hitches with the warp ends directed toward the interior of the rug. I then worked the warp ends back into the woven edge of the rug with a tapestry needle. I worked the ends into the rug for about half to three quarters of an inch and brought the needle back out to the surface. By alternating the places where the needle exits, it doesn't create a ridge and the finished edge will lay flatter. I pressed the finished rug with a steam iron.

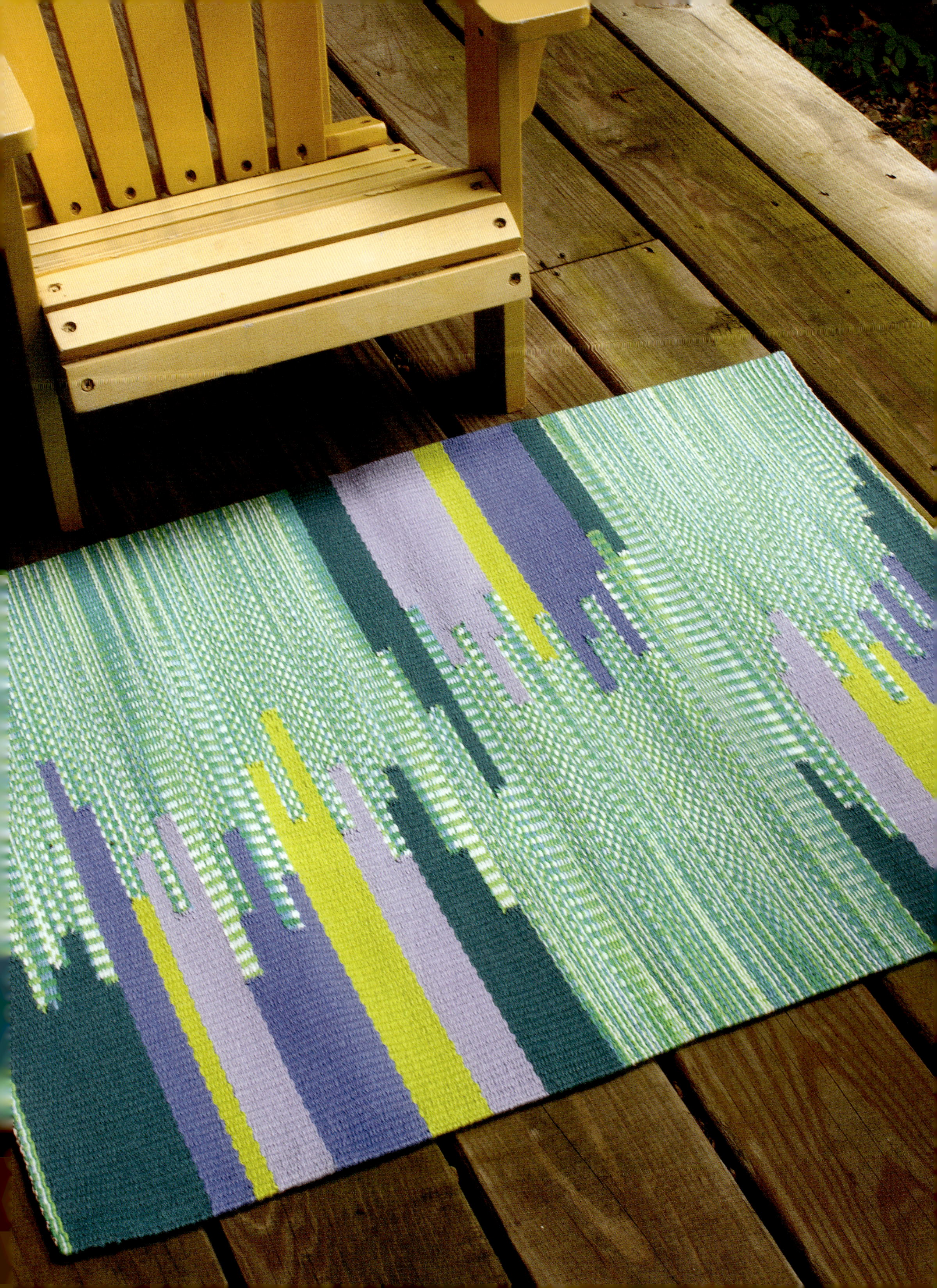

4				4				4			
	3				3				3		
		2				2				2	
			1				1				1

	4
3	
	2
1	
X	
	X
X	
	X

Warp

12/6 cotton seine twine (Fishgarn): Natural

Total ends: 200

Sett: 8 epi (4 working epi); double the ends in the dent and heddle eye; use an 8-dent reed and sley every other dent

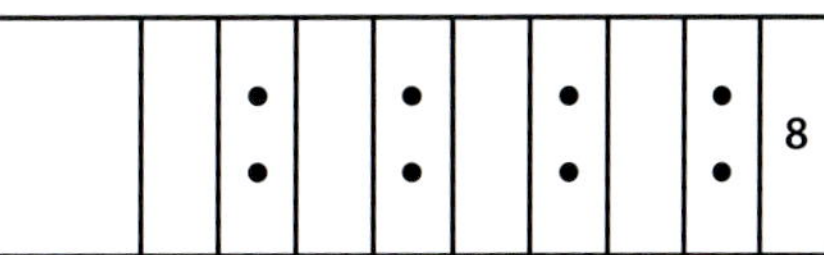

Width: 25 inches

Weft

Sugar'n Cream cotton yarn: Emerald Energy Ombre, Hot Green, Cornflower, Blueberry, and Teal

Volcano Rug

It has been my experience that working with three colors in a single row of clasped weft technique can have some challenges. I wove several samples until I found one that I liked the best. You see, when working with three colors, there is a time when the shuttle that carries the center color must come up and out of the warp so that you can change the shed and then reenter the warp in the exact same place.

I started out by putting the Cayenne red color on the floor on the left side of my loom. I placed the Stone color on the right side of the loom. I planned for the Black color to weave in the center of the rug. I started the pattern by opening the 1–3 shed and throwing the shuttle from right to left. I then clasped the Cayenne thread and pulled the clasped threads into the shed from the left and placed the join where I wanted it to fall and let the shuttle exit out on the right side. I closed the shed and beat the threads down. I then reopened the 1–3 shed and let the shuttle clasp the Stone thread on the right and reentered the shuttle back into the shed. I brought the shuttle up and out of the web somewhere in the middle of the warp and laid the shuttle down on the beginning woven area of the rug. I pulled on the Black thread to place the Stone thread and this new join somewhere in from the right selvedge. I reopened the shed to allow me to tuck the beginning Cayenne and Stone tails of thread into the open shed and beat them down. It was now time to change to the next shed.

I opened the opposite plain weave shed (2–4) and reentered with the shuttle where it had just exited. I pushed the shuttle to the left and clasped the Cayenne thread. I then threw the shuttle to the right selvedge and pulled the Cayenne thread and the clasped join into the shed and placed the join where I wanted it to fall. I closed the shed and beat. Then, as before, I reopened the 2–4 shed and took the shuttle around the Stone thread and pushed it back in the shed from the right. I pulled the Black thread until the new join was where I wanted, closed the shed, and beat.

You will notice that at the spot where the shuttle exits and enters, there appears a float that can't be avoided. It is the nature of this change from one shed to the other shed at this point. Using the black color in the center helped to camouflage the float better than the other colors used in this center section. I wove several samples with the colors in different positions, and this option seemed to work the best for me.

While working the Three-Colored Towel warp (page 37), I discovered a better way to exit the web and then reenter the shuttle into the next shed without creating such a noticeable change. I brought the shuttle out at the spot where the clasped join was being placed on the right side of the work. I then changed the shed and reentered the shuttle at this spot. It appears cleaner and much neater to my eye because I was working with a much finer thread. The heavy cotton accentuated the float.

I continued this way until I had reached my desired length. When I was finished, I cut the three colors, leaving tails that were several inches long. I then reopened the shed and worked the tails back into the shed and beat the tails down to hold them in place. After weaving the hem area, I cut the tails so they were flush with the surface of the rug.

4				4				4			
	3				3				3		
		2				2				2	
			1				1				1

				T	T	
		4	4		4	
	3	3		3		
2	2				2	
1			1	1		
				X		Hem
					X	
				X		
					X	
X						Body of Rug
		X				
X						
		X				
				X		Hem
					X	
				X		
					X	

Warp

Berga 12/6 cotton seine twine: Dark Brown

Total ends: 200

Sett: 8 epi

Width: 25 inches

Weft

8/4 cotton for hems: Black (doubled)

Mop cotton: Black, Stone, and Cayenne

Wearables

Purple Rain Scarf

I like to support independent dyers. I often wonder what they might have been imagining when they dyed certain colors together in a skein of yarn. I do a little dyeing myself, so I know from experience what color combinations are pleasing to my eyes, and I have learned what colors NOT to place beside one another. Sometimes you get a skein that doesn't turn out to be very pretty, but don't despair—you can always overdye it in a whole new colorway, so all is not lost in your adventure.

I look to nature and art for some of my ideas. When spring starts to burst with flowers and blossoms, I get excited about trying to reproduce these colors in a skein of yarn. It's never the same, but it's always pretty. As in art, sometimes a photograph or painting will be the inspiration for a new colorway. Good ideas are all around us, and we just need to stay open to the next inspiration.

I found two skeins of 10/2 cotton yarn that were dyed by an independent dyer called Crabapple Yarns. One skein was dyed with soft pinks, and the other skein was dyed with deep purples, lavender, and reds. I liked how the two skeins looked together, so I decided to incorporate them into a clasped weft scarf. I chose 8/2 Tencel yarn for the warp in a rich purple.

This was my first attempt in weaving clasped weft in a twill pattern. Since I wasn't familiar with how to handle the selvedges in a twill weave, I decided to play it safe and thread for a plain weave selvedge. This approach would require two additional shafts to be used in the threading of the scarf. Look at the threading draft and tie-up to see how I threaded the warp and wove it. I didn't have any problem controlling the edges as I wove. I thought about using a floating selvedge, but trying to remember the over/under sequence order and the clasping of the wefts was more than I wanted to tackle.

I like the look of long, diagonal runs of color. I wove moving the clasps slowly in one direction, and when the spirit moved me, I moved the clasps back in the opposite direction. That's the beauty of weaving this way. You can have quick movement changes that resemble an EKG readout, or you can move the clasps more gradually, as I did in this piece. The choice is yours. Perhaps try both to see how you like it.

I finished the ends with a simple hemstitch to hold the edge and then twisted the fringes for a pretty look.

Border				12X																Border			
6		6																		6		6	
	5		5																		5		5
						4				4		4				4							
					3				3				3				3						
				2				2						2				2					
							1				1				1				1				

				T	T	
	6		6		6	
5		5		5		
		4	4		4	
	3	3		3		
2	2				2	
1			1	1		
				X		Hem
					X	
				X		
					X	
X						
	X					
		X				
			X			
X						
	X					
		X				
			X			
X						
			X			
		X				
	X					
X						
			X			
		X				
	X					
				X		Hem
					X	
				X		
					X	

Warp

8/2 Tencel: Iris

Total ends: 200

Sett: 20 epi

Width: 10 inches

Weft

Crabapple Yarns 10/2 hand-dyed cotton, 1 skein dark purples and 1 skein light pinks

Note:

This special border threading made a neat selvedge. You can also omit this border and weave on a 4-shaft loom without needing a floating selvedge.

Naturally Neutral Scarf

Tencel, like rayon, is made from cellulose fibers. The properties of this thread make for a garment with drape and an extremely comfortable hand when you touch it. It feels very much like silk, and it has the shine and luster of silk too.

I chose to go with an asymmetrical approach to the design of this scarf. The color change in the warp is off center to add interest.

While weaving, I moved the weft colors with every woven pick across the centerline and into the opposing color field. That is, when the Black and Natural wefts met to clasp, I moved the Natural weft thread into the Black field to where I wanted it to fall and then squeezed it into place. I used a very light beat so as not to make the fabric too stiff. In the next pick, I moved the clasped join so that the Black weft crossed over into the Natural field and again beat very lightly. Back and forth I wove, making sure that the clasped joins never stacked on top of each other in the same place. This makes them appear random with a jagged edge.

For the beginning and ending hemstitched edges, I used Natural thread on the Natural areas and Black thread on the Black areas. As I sat at the loom to weave this piece, the Natural-colored warp threads were on my left and the Black warp threads were on my right. To start, I left a long tail of Natural thread on the left selvedge edge to act as my stitching thread and a long end of Black thread on the right. I wove about 1 inch into the scarf and then hemstitched from the right with the Black thread until I got to the color change. I worked the remaining Black thread back into the woven edge to secure it. I then hemstitched from left to right with the Natural thread to the color change and again worked the remaining stitching thread back into the woven hem. I used four warp ends in each of the hemstitched bundles.

Finish your scarf by hand washing it in warm water with a little mild soap. Rinse and then soak your scarf in warm water with a little unscented fabric softener. The softener will improve the drape of the fabric. Rinse and then air dry. Iron on a rayon setting.

	4
3	
	2
1	
X	
	X
X	
	X

Natural				Black			
4				4			
	3				3		
		2				2	
			1				1
30X 120 Ends				15X 60 Ends			

Warp

8/2 Tencel: Natural, 120 ends; and Black, 60 ends

Total ends: 180

Sett: 20 epi

Width: 9 inches

Weft

8/2 Tencel: Natural and Black

Cityscape Scarf

Here is an example of how to use up those small amounts of yarn in your stash. I simply picked out several colors of Tencel thread that I thought looked good together. There wasn't enough left on any one cone to weave a whole project, so I combined them by making stripes of different widths. This approach added a lot of interest to the warp. I am a big fan of asymmetrical designs. I must be honest with you, though: Much of my weaving is balanced top to bottom and left to right. When I get the chance to stretch out and live on the edge, so to speak, it feels good and almost naughty. It feels like I am getting away with something that I know is acceptable but frowned on by those "traditional weavers."

When I started out to wind the warp, I thought I would start the design with the cooler, blue colors on the left and wind toward the middle of the warp, and then switch over to warmer colors for the remaining right side of the warp. I had a certain number of ends in mind to make a scarf warp, but if it ran a little short or a little wider, that would be all right with me. Luckily for me, I had plenty of thread to make the scarf and had some leftovers of each color.

I threaded the loom for a straight draw twill—nothing fancy, but a threading that would give me plain weave.

The colors for the weft were colors that were left over from making the warp. I had quite a lot of blue and cream remaining, so these were the chosen two. I wove with the blue cone on the floor to my left, and I introduced the cream color on a shuttle from my right. The treadling was a simple plain weave. As I wove the odd picks (1 and 3), I dragged the blue weft across to the right side of the warp where the warm-colored stripes were. When I wove the even picks (2 and 4), I pulled the clasped ends over a little and placed the join somewhere in the cooler-colored blue stripes. I continued this way, going back and forth, throughout the entire weaving of the scarf.

After finishing the ends, I hand washed the scarf in warm, soapy water and then put a small amount of unscented fabric softener in the rinse water. I hung it to dry and then gave it a good pressing.

4				4				4			
	3				3				3		
		2				2				2	
			1				1				1

Warp Color Order and Number of Ends

							16	Black
						16		Lemon Drop
					20			Dark Teal
				20				Spice
			40					Lemongrass
		40						Cerulean Blue
	40							Royal
40								Amethyst

	4
3	
	2
1	
X	
	X
X	
	X

Warp

8/2 Tencel: Black, Lemon Drop, Dark Teal, Spice, Lemongrass, Cerulean Blue, Royal, and Amethyst (see color chart for order and number of ends)

Total ends: 232

Sett: 20 epi

Width: 11.5 inches

Weft

8/2 Tencel: Cerulean Blue and Lemon Drop

Bink's Scarf

I love designing striped warps, and I especially love stripes that have a gradient look. The gradient width stripes allow the colors on either side of the warp to appear as if they are slowly bleeding into one another. I first took notice of this design years ago when I saw a custom-painted car that had one color dominating the front fender and a different color painted on the rear fender. The forward color gently moved toward the mid-side panel door of the car in a series of stripes that were graduated from wide to narrow. Between each wide stripe of the front color, there was another stripe made up of the rear fender's color. That color went from very narrow to wider and wider as the front color became narrower. It was striking, and it made such an impression on me that I have used it many times to add interest to a possibly uninteresting piece. Notice how I used it in the two shawl warps in this book (pages 76 and 79).

For this scarf, I thought I would push the design a little further. Instead of using just two colors for the gradient (light and dark), I used two different, darker blues for what would make up the dark side of the scarf and two lighter colors—soft yellow and lime green—for the opposing area. I carefully alternated the two dark and two light colors as I sleyed the reed and threaded the heddles. Refer to the draft for the details.

For the weft, I used a single light thread and a single dark thread to weave and do the clasping. The interest is clearly in the warp planning.

I finished the scarf with a hemstitched edge and cut the fringe to a short one-inch length. I cut the fringe after the scarf was wet finished. To do this, I hand washed the scarf in warm water with a little soap and fabric softener. I like using the unscented variety of fabric softener. I hung it to dry and gave it a light pressing with an iron to remove the wrinkles.

4				4				4			
	3				3				3		
		2				2				2	
			1				1				1

Warp Color Order

Ends																																									
55				1				1				1				1				1				1				1				1				1				1	Blue Ming
55			1				1				1				1				1				1				1				1				1				1		Royal
55		1				1				1				1				1				1				1				1				1				1			Lemon Drop
55	1				1				1				1				1				1				1				1				1				1				Lemongrass
Total Ends 220																																									
	10X		1X		9X		2X		8X		3X		7X		4X		6X		5X		5X		6X		4X		7X		3X		8X		2X		9X		1X		10X		

	4
3	
	2
1	
X	
	X
X	
	X

Warp

8/2 Tencel: Blue Ming, Royal, Lemon Drop, and Lemongrass

Total ends: 220

Sett: 20 epi

Width: 11 inches

Weft

8/2 Tencel: Royal and Lemongrass

Blueberries and Grapes Shawl

I have used this cotton bouclé thread many times as weft in dish towels. It gives a nice texture and weight to a towel. I like it so much that I wanted to weave a shawl with it as both warp and weft. It wove nicely without a single broken warp end. The clasped weft doubled the weight of the weft yarn, making it very nice as shawl fabric. It's perfect for those cooler evenings outside and in heavily air-conditioned restaurants. I like the color combination of Blue, Lime, and Magenta. Of course, please change out the colors to suit your favorite color palette. Notice in the warp that there is one Black thread between each color change. I found this arrangement helped to define the stripes.

When I wove the fabric, I clasped the wefts within the center of the Lime-colored stripe and at the edge of the Blue or Magenta stripe (see photos). I made sure that the clasped join always fell exactly on the edge of the stripe. I alternated the weft colors, pick for pick, giving it a subtle pattern.

Like the Tencel Naturally Neutral Scarf pattern (page 67), I finished the edge with a hemstitch that used the Magenta color on the Magenta side of the shawl and the Blue stitching thread on the Blue side of the shawl. I worked the stitches from left to right and then from right to left. I worked the ends back into the woven hem. I then twisted the warp ends to keep the fringe ends tidy and so they wouldn't get tangled. I hand washed the shawl in warm soapy water and rinsed it in warm water. I then gave it a soak in warm water with unscented fabric softener for about an hour before I rinsed out the softener. I let it air dry and gave it a light pressing with an iron on a cotton setting.

4			
	3		
		2	
			1

Warp Color Order

	Ends																																
	72																	2		4		6		8		10		12		14		16	Blue
	128		2		4		6		5		10		12		14		16		14		12		10		8		6		4		2		Lime
	72	16		14		12		10		8		6		4		2																	Magenta
	30																																Black: insert 1 Black thread between each color change
Total Ends	**302**																																

	4
3	
	2
1	
X	
	X
X	
	X

Warp

Brassard Cotton Bouclé: Blue, Lime, Magenta, and Black

Total ends: 302

Sett: 12 epi

Width: 25 inches

Weft

Brassard Cotton Bouclé: Blue and Magenta

Autumn Snuggles Shawl

My dear friend and extraordinary photographer, Kathy Eckhaus, gravitates to the colors of fall. When I thought about doing a wool shawl with gradient stripes, I had Kathy in mind and used the colors Cypress, Mustard, and Russet.

The warp and weft yarns to weave this shawl are from Harrisville Designs. Harrisville yarns are the perfect choice for this shawl because Harrisville spins two different weights of wool yarn. Their Highland weight yarn is what I used for the warp. It has 900 yards to the pound. The weft yarn for this shawl is Harrisville's Shetland style and is conveniently half the weight of the warp yarn. Shetland has 1,800 yards to the pound. It is also convenient that all their colors are available in both sizes of yarn. When the Shetland yarn is doubled in the weft direction, the weft is now the same size as the warp yarn, producing a balanced fabric.

When weaving this shawl, I let the clasps fall anywhere within the center stripe. There is a subtle zigzag appearance within the center stripe. This approach to the clasping technique is different from the cotton bouclé shawl. The bouclé shawl has a very controlled approach to the placement of the joins, while in this piece the spacing is very random.

Like the Cityscape Scarf (page 70) and Blueberries and Grapes Shawl (page 76), I finished the ends with two different colors of hemstitching thread so that the stitches blend in visually with the shawl's edge (see photo). Please reference these other projects to see the hemstitching. I hand washed the wool shawl in hot water with a mild shampoo with hair conditioner and let it soak for approximately half an hour to allow the yarns to bloom. I then rinsed the shawl in warm water and spun out the excess water in the washer on the spin cycle only. Air dry your shawl, and either press and smooth out the surface with your hand or iron it on a low, wool setting.

4				4				4			
	3				3				3		
		2				2				2	
			1				1				1

Warp Color Order

	Ends																												
	56	14		12		10		8		6		4		2															Cypress
	98		2		4		6		8		10		12		14		12		10		8		6		4		2		Mustard
	56															2		4		6		8		10		12		14	Russet
Total Ends	210																												

	4
3	
	2
1	
X	
	X
X	
	X

Warp

Harrisville Designs Highland: Cypress, 56 ends; Mustard, 98 ends; and Russet, 56 ends

Total ends: 210

Sett: 8 epi

Width: 26.25 inches

Weft

Harrisville Designs Shetland: Cypress and Russet

I used two different colors of hemstitching thread

Table Runners

A Quake's a Coming! Table Runners

When you look at this table runner, does it remind you of a seismograph reading? It does me. I find the back-and-forth, sporadic movement of this design exciting. I look at the runner and hear Carole King singing, "I feel the earth move, under my feet. I feel the walls tumbling down, tumbling down!"

This is a great starter project if you are trying clasped weft for the first time. The warp is one color, and it's woven as plain weave. The magic happens with your yarn choices. I found two variegated cotton yarns called Araucania Yarns. One yarn is made up of pinks and charcoal colors, and the other skein has warm colors of squash and pumpkin. I liked the way the two skeins looked together, so I decided to try weaving them in a rather sporadic, back-and-forth manner to see what it might look like. I loved it. Never mind that the runner doesn't have colors that go in our home. We have white dishes, and it looks great down the center of our table.

Just for fun, I wove a second runner with a bit more planning. I started with the pink/charcoal on the left and the pumpkin/squash on the right. I wove that way for a section and then wove with both yarns in the same shed together to make a divider. Doing so allowed me to switch sides and weave with the pumpkin/squash color on the left side and the pink/charcoal color on the right. I then wove the double stranded divider again and then switched sides again. I continued this way until the runner was as long as I wanted. Look how different these runners are from each other.

This is a great way to help reduce your stash of yarns. Simply find two contrasting skeins of yarn you like and use them for your weft. I chose two variegated yarns, but you could use two solid-colored yarns or even weave with one solid color yarn and one variegated to give you a look like some of the towel examples in this book.

4				4				4			
	3				3				3		
		2				2				2	
			1				1				1

	4	Basket Weave Tie-up
	3	
2		
1		
X		Hem Color Rouge Vin
	X	
X		
	X	
X		Body of Runner
	X	
X		
	X	
X		Hem Color Rouge Vin
	X	
X		
	X	

Warp

8/2 cotton: Rouge Vin

Total ends: 320

Sett: 16 epi

Width: 20 inches

Weft

8/2 cotton: Rouge Vin (for hems)

Araucania Yarns Variegated Cotton: pink/green/charcoal and squash/pumpkin

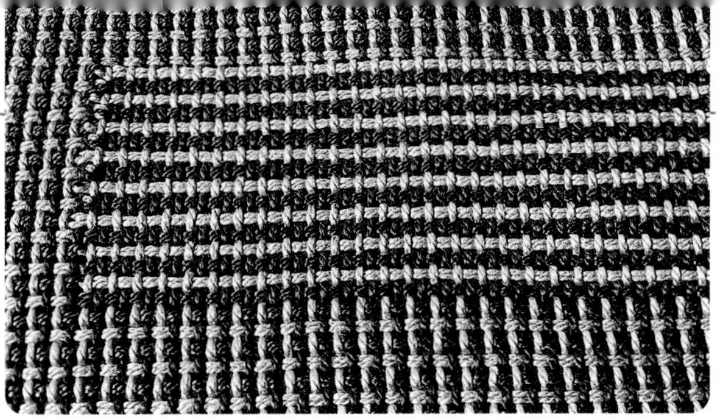

Optical Illusion Table Runner

The log cabin pattern is most likely the best-known example of color and weave effects. This result happens when two contrasting colors, light and dark, alternate in the warp-way direction and are then woven with alternating light and dark threads in the weft direction.

The traditional log cabin pattern is woven as plain weave and has two pattern blocks. The pattern blocks are achieved by threading the first pattern block with the dark threads on the odd shafts and the light-colored threads on the even shafts. This setup is repeated for a desired number of times, and when you wish to change to the next pattern block, you change the order of the color arrangement. Now the light-colored threads are threaded on the odd shafts and the dark on the even-numbered shafts. This shift to the other pattern block places two light threads side by side. When you switch back to thread the first pattern block arrangement, you will notice that two dark warp threads are now together. You thread the loom alternating the blocks all across the width of the warp.

The treadling order also alternates light and dark colors, often following the same order as the warp in what is called "tromp as writ." This translates to "Treadle as it is written in the threading draft." This is often the case, but not always, and you can treadle it any way that you like. It does need to be treadled as plain weave.

When I thought about clasped weft in a color and weave combination, it became apparent that I could simply thread all the dark-colored threads on the odd-numbered shafts and the light on the even-numbered shafts, all the way across the warp. No switching to make blocks. The weft arrangement was another story. To weave this runner with its freeform blocks, I would need to have two shuttles on the right side of the loom, one with light thread and one with dark thread. I would also need two cones, one of the light and the other of the dark thread, on the left side of the loom.

To weave the pattern, you will first open the shed and weave the dark shuttle across the warp to clasp the light-colored thread and drag it back into the shed where you wish the pattern block to start. Beat it down gently and change the shed. Now throw the light-colored thread across in this new shed and clasp the thread coming from the dark cone. Move these threads and their clasped join into the shed and place the join right above the last pick. Continue this way until you have built up the pattern block to a height that you like.

To make a new pattern block, change the color sequence in your weft. This will always put two light threads together at the switch or two dark threads, just as it does in traditional log cabin. Your treadling will always follow plain weave treadling. You are at the controls as to where you want your pattern blocks to be located. Remember, there is no wrong way here.

One note of caution: Be mindful of the wrapping of two shuttles on the right side of your work, and you will notice there is some tangling that happens on your left side where the cones sit on the floor. I found it helpful to stop periodically and untwist the cones. It's a slow process to weave, but it has so many possibilities.

I enjoyed the freedom to just "wing it" and place the clasped joins wherever I pleased without any preconceived thoughts or draft to follow. I think you are going to enjoy trying this project.

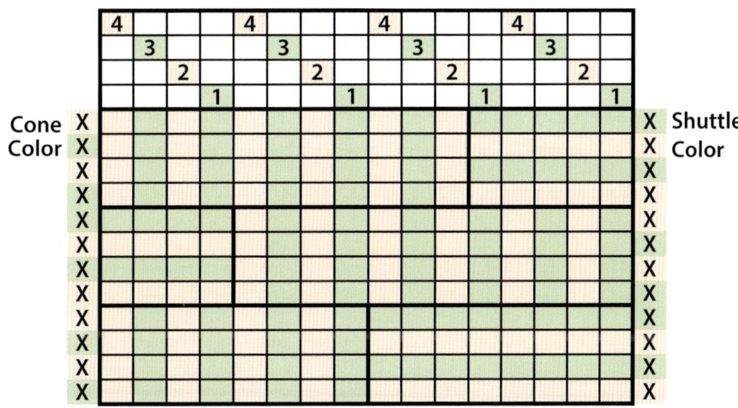

	4
3	
	2
1	
X	
	X
X	
	X
X	
	X
X	
	X
X	
	X
X	
	X

Warp

3/2 perle cotton: Desert Sand and Loden Green

Total ends: 240

Sett: 12 epi

Width: 20 inches

Weft

5/2 perle cotton: Desert Sand and Loden Green

Morning Has Broken Table Runner

Overshot is a favorite structure of mine to weave. Weaving overshot was one of my first adventures when I got my four-shaft loom. There are hundreds of pattern possibilities that can be woven in overshot, and I like to kid myself and think that I will someday weave them all (or at least memorize all their names). I do have a few patterns that I seem to revisit again and again. One pattern I use a lot is called Star of Bethlehem. The diamond shape of this pattern is very twill-like. Perhaps that is why I like it.

Let's take a moment to understand how the weave structure of overshot works. There is one warp that is threaded to the overshot pattern. Overshot is considered a twill derivative. I think of it as a twill hybrid. The weft pattern is made by weaving with two shuttles. One shuttle carries a bobbin with a pattern thread. The pattern thread is usually twice the size of the warp thread so that it is more pronounced. Then there is a second shuttle required that carries a tabby thread. This tabby thread is usually the same size thread or a little finer than the warp thread. When weaving overshot, the pattern shuttle leads in the weaving. The weaver follows a written pattern. Each pattern pick will be any one of the four standard twill combinations. The pattern picks can be repeated as the pattern suggests. This is made possible because there is a tabby thread woven between each pattern pick that is not written into the pattern. The draft will indicate that a tabby should be used. The two tabby combinations alternate between the pattern picks, weaving a plain weave ground cloth. If the pattern threads were removed from a piece of overshot, there would still be a usable piece of cloth because of this plain weave ground cloth. So overshot can be described as a structure with one warp and two wefts. One weft weaves a plain weave ground cloth, and the pattern weft is a supplementary twill weave.

For this table runner, I used the overshot pattern Star of Bethlehem. I used a rosepath twill for the border threading. To continue my exploration in clasped weft, I decided to clasp the tabby weft. This means you will see the clasp as part of the ground cloth. This approach would require a finer tabby weft than I would normally use for overshot because the clasping of the two weft threads would now double the tabby weft. Since the warp was threaded with an 8/2 cotton, I used 16/2 cotton for the tabby wefts in two colors. This choice made the weaving a little tricky. I wove the pattern weft with 5/2 cotton as I usually do, and for the tabby wefts, I placed a cone of Natural 16/2 cotton on the floor to my left. I used a boat shuttle with the darker tabby attached on my right side. After weaving a pattern weft, I opened the tabby shed and threw the dark tabby into the open shed to the left. I picked up the light cotton thread coming up from the floor and clasped the two threads. I then threw the shuttle back to the right side, pulling the light thread into the open shed. When I decided where the clasp should fall, I closed the shed and beat it into place. I then wove another pattern row and repeated the clasping tabby in the opposite plain weave shed. I continued weaving this way for the length of the runner.

I finished the piece with a traditional rolled hem. Please note that the hem also has been woven with a clasped weft to maintain uniformity.

		4								4		4		4				4		4				4		4		4								4			
	3						3		3		3						3				3						3		3		3						3		
2				2		2		2								2						2								2		2		2				2	
			1		1								1		1				1				1		1								1		1				1

8X

		4		4			
	3				3		
2						2	
			1				1

3X

Repeat 3X at the beginning and end of the threading

				T	T	
		4	4		4	
	3	3		3		
2	2				2	
1			1	1		
				X		Hem
					X	
X						Beginning Border 3X Use Tabby
	X					
		X				
			X			
X						
			X			
		X				
	X					
X						Pattern Area Repeat to Desired Length Use Tabby
	X					
		X				
			X			
X						
	X					
		X				
			X			
X						
	X					
		X				
			X			
		X				
	X					
X						
			X			
		X				
	X					
X						
			X			
		X				
	X					
X						Ending Border 3X Use Tabby
	X					
		X				
			X			
X						
			X			
		X				
	X					
				X		Hem
					X	

Warp

8/2 cotton: Natural

Total ends: 368

Sett: 20 epi

Width: 18.5 inches

Weft

Pattern thread: 5/2 perle cotton in Magenta

Tabby: 16/2 cotton in Natural and Black

Puzzle Pieces Table Runner

Can you see how the colorful blocks resemble the pieces of a jigsaw puzzle? I was inspired by a room divider that used these shapes in the design of the gridwork. I thought it was an unusual architectural feature and wanted to use it somehow in a woven project. I thought the interlocking pieces would work well for a clasped weft design, and you can see they did indeed.

As I laid out the pattern on a piece of graph paper, it quickly occurred to me that each interlocking piece could be assigned a different color. I chose to use the six colors on a color wheel that make up the primary and secondary colors: red, orange, yellow, green, blue, and violet.

For the warp threads, I chose to use Black 8/4 cotton carpet warp. The warp is 24 inches wide in the reed. As part of the design, I divided the warp into four sections. Each section has an equal width of 6 inches. The threading is a broken twill. Take a look at the threading draft to get a better understanding of the warp's layout. The visual breaks in the threading helped a great deal in placing the clasped joins. I could easily see where to place the joins to stack them on top of each other. I wove row after row, being careful to line up the clasps vertically so the colors met with a clean vertical line. When I was ready to make a change in the design, it was easy to see where I needed to move the clasped joins because of the breaks in the twill direction.

I chose the colors of the rainbow thinking this would look great on a table set with Fiestaware plates and cups. Please change out the colors to suit your taste.

	4			4					4			4			
3					3			3					3		
			2			2					2			2	
		1					1			1					1
18X 72 Ends				18X 72 Ends				18X 72 Ends				18X 72 Ends			

				T	T	
		4	4		4	
	3	3		3		
2	2				2	
1			1	1		
				X		Hem
					X	
				X		
					X	
X						Body of Runner
	X					
		X				
			X			
				X		Hem
					X	
				X		
					X	

Warp

8/4 cotton carpet warp: Black

Total ends: 288

Sett: 12 epi

Width: 24 inches

Weft

8/4 cotton carpet warp: Grass, Cranberry, Lime, Moody Blue, Eggplant, and Burnt Orange

Note: *Each block on the graph is equal to 12 clasped picks.*

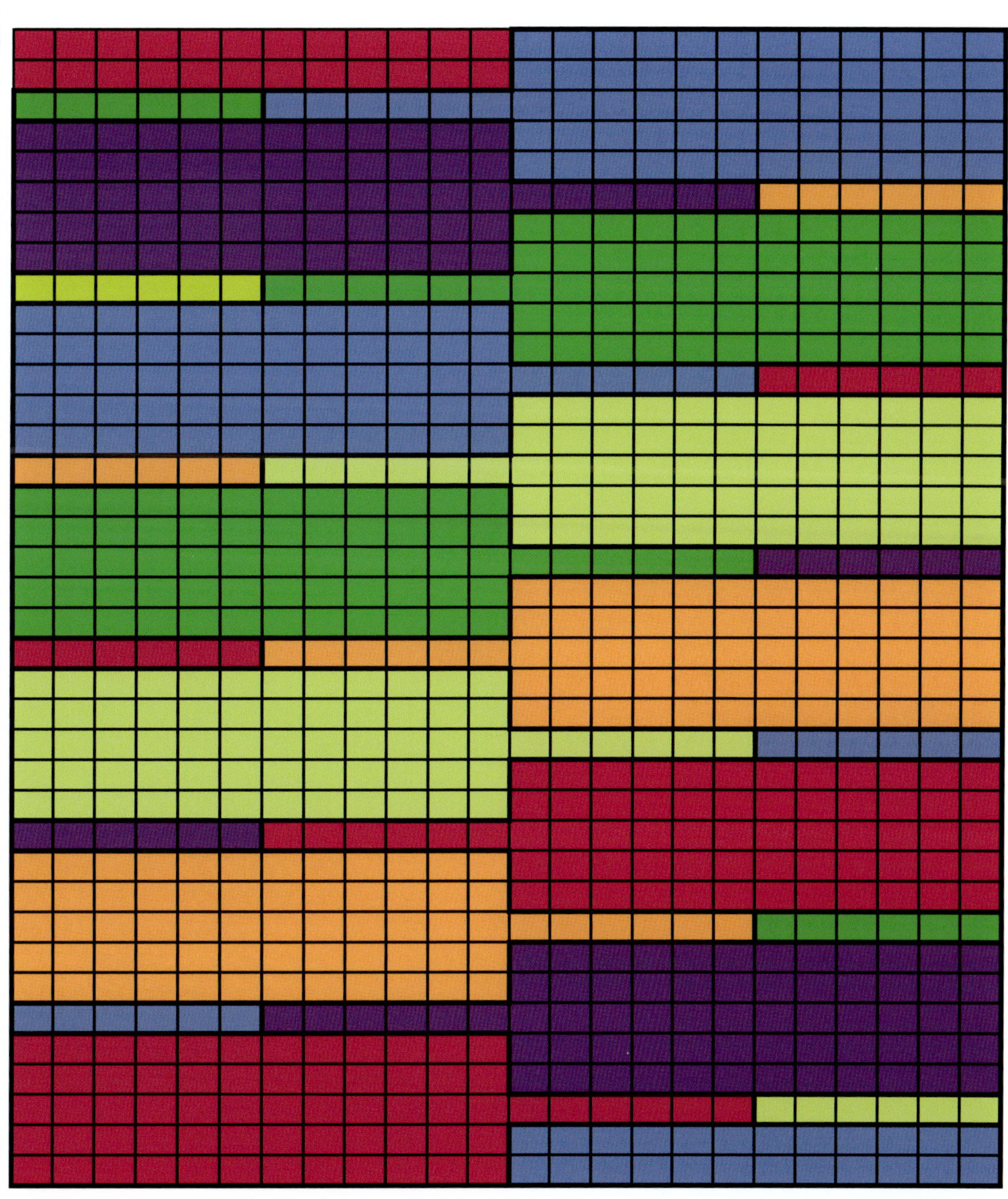

Bits and
Bobs

Weaver's Secret Baby Blanket

What baby wouldn't love this cheery blanket? Of course, you can always change out the colors and make it your own.

I like the look of a simple four-shaft broken twill. My wife and I use this option all the time when designing dish towels because we like the subtle visual break in the threading where it makes a color change.

I thought this might be a good choice to use for clasped weft. The clasps can be made at those color breaks to make a clean vertical line in the pattern. Another thought I had about the weaving of this blanket was: If I use 8/2 cotton for the weft, when I clasp this thread, it would now be the size of 8/4 cotton. If I double the 8/2 cotton in the warp by sleying two ends in a dent and then thread these two ends in the heddle, the fabric should be balanced when it's woven. That is what I did, and it *is* beautifully balanced.

My cat "helping" with the draft for the baby blanket

The pattern started out with the red color on the left and the blue weft thread on the right. The clasping was done by having the two weft colors meet at a color break in the warp. The treadling follows a twill sequence and is woven until the color blocks are square. This is where the clasping colors end and the gold color is introduced. The gold color weaves across the entire width of the warp from selvedge to selvedge. Since the wefts in the clasped areas are doubled, you will want to double the gold thread in the shed as well. Toss the shuttle across the warp and catch a selvedge thread with your gold thread. Then toss it back in the same shed so it now becomes a doubled weft thread. (Or, if you have a shuttle that will accommodate two bobbins, you will need to throw only one time per shed.) This keeps the fabric balanced. Weave the gold area to square as well. For the next clasped sequence of red and blue, exchange sides and weave with the red on the right and the blue on the left. Again, clasp the joins at a warp color break but different from the last area. Weave the blocks to square. This process is known as weaving tromp as writ—weaving in the same order that the threading draft is written. Follow this up with another gold section. Repeat six times and balance the ending blocks the same as in the beginning.

Roll a hem to secure the ends. Wash in warm water and tumble dry in the dryer. The result is a soft and snuggly blanket that any baby will love. I named this blanket Weaver's Secret because only a weaver will know how this design was made!

Double bobbin shuttle

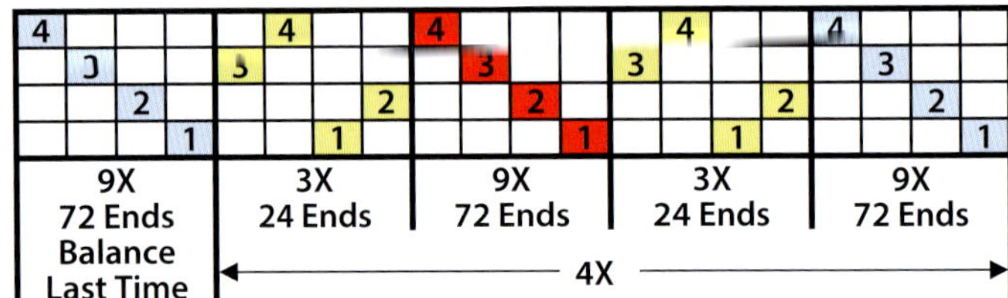

Note:
Weave blocks to square.

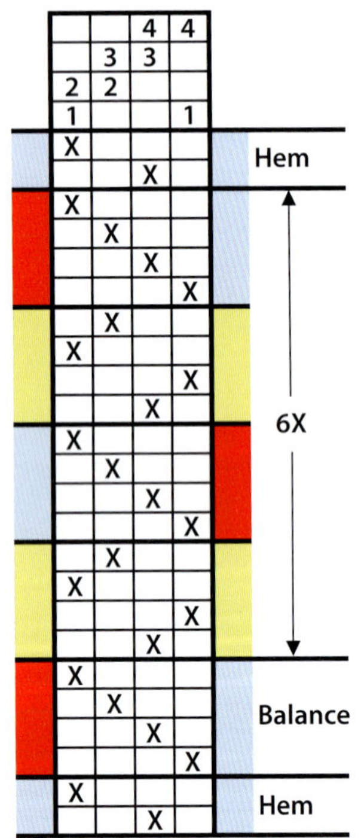

Warp

8/2 cotton: Jeans Blue, 360 ends; Viel Orange, 192 ends; Cayenne, 288 ends

Total ends: 840

Sett: 24 epi (12 working epi). Sley 2 per dent in a 12-dent reed. Thread 2 ends in a heddle eye as if they were one end.

Width: 35 inches

Weft

8/2 cotton: Jeans Blue, Viel Orange, and Cayenne

Ocean View Pillow

When my wife and I travel, we like to see whether there are any yarn shops in the area. We found a lovely shop in Kodiak, Alaska, called The Rookery. My eyes lit up when I saw the large selection of Noro yarns. These are beautiful yarns from Japan that are quite unique. The skeins are known to have long runs of a color before they gradually change into another color for a striking, gradient look. Each color may be as long as a yard in length, but there is a slight blending of colors where two colors overlap. Because of the unique way the colors are arranged, it is important to remember to wind a bobbin until full and then rewind it so that the colors stay in their correct order.

I started out by winding two skeins into separate balls on a ball winder. One ball could stay as it was because it would be sitting in a coffee can on the floor to the left of the loom as a constant feed into the open shed without any break in the color sequence. The second skein would be wound onto a bobbin and placed in a boat shuttle. These bobbins had to be wound and then rewound to keep the consistent flow of color.

I didn't jump into this project right away. I had to ponder all the possibilities before getting started with the weaving. The colors in the two skeins look very similar to the colors in our living room. Why not weave fabric for a pillow? I seriously thought about a warp of neutral colors such as black, charcoal, beige, and brown. They looked all right, but all right was not good enough. I decided to go with a soft, pale blue 10/2 cotton and threaded the loom to an expanded point twill. This approach resulted in a subtle pattern for the background while the Noro yarns were the main focus of the fabric's overall look.

I threaded the warp for a 20-inch width to allow plenty of room for take-up and shrinkage when the fabric was washed and wet finished. The fabric was machine washed on a gentle setting in warm water. It was then machine dried on a regular setting. I wove about 48 inches of fabric so that I could select a small portion to be used from this long length of woven fabric. You might think that this is a little wasteful, but you can be sure that the remaining scraps of this gorgeous fabric will be used in something else.

Our original thought was to buy a pillow form, zipper, and a solid-color fabric for the back of the pillow. Then my thrifty and quick-witted wife thought of looking at a home accessory store for a pillow that could be reworked to accommodate the fabric. Cindy found an 18-inch pillow that was the perfect match on her first try, and the best part of the story is that the pillow was on closeout price. Cindy took the pillow apart, saving the back portion and the pillow form. She then sewed the handwoven fabric to the front of the pillow. We now have a pretty new pillow on our sofa that reminds us of the wonderful time we had in that yarn shop in Kodiak, Alaska.

		4				4				4		4				4				4			
	3				3				3				3				3				3		
2				2				2						2				2				2	
			1				1				1				1				1				1

20X

				T	T	
		4	4		4	
	3	3		3		
2	2				2	
1			1	1		
				X		Plain
					X	Weave
X						
	X					Twill
		X				
			X			

Warp

10/2 cotton: Periwinkle

Total ends: 480

Sett: 24 epi

Width: 20 inches

Weft

Noro Taiyo Sock Yarn: colors S-3 and S-8

Reykjavik Placemats

I have a longtime friend who was a weaver, but at some point she gave up her looms for a rug hooking frame and piles of wool fabric. She is now a renowned rug hooker and spends most days pulling some loops. She went to Iceland with a like-minded group of fiber artists and fell in love with that beautiful country. When she returned, she wanted to hook a rug to remind her of the trip. She was enamored with the signs that preceded each town and thought they were unique to Iceland and would be a perfect reminder. The signs are black and yellow with the name of the town underneath the building silhouettes. If she omitted a town name, this would be a rug that reminded her of any city, anywhere in beautiful Iceland.

When I saw her rug, I thought it would be fun to weave some matching placemats, so I took a picture of the sign and transferred it to graph paper. The scale I used was one square to represent one square inch on my mat. I put the Black yarn on the floor to the left of my loom and then used my chart to determine how far to pull it with the Yellow. Some of my rooflines got a little wonky, but there is a lot of seismic activity in Iceland, so perhaps they were experiencing a tremor! At any rate, I love the matching set of placemats and floor rug, and I hope my friend will have lots of dinner parties with an Icelandic theme.

This same technique could be used to weave the silhouette of the buildings on your family homestead, your favorite place of worship, or even the castle you dreamed you would own someday. Get out your graph paper and have fun.

4				4				4			
	3				3				3		
		2				2				2	
			1				1				1

	4	
3		
	2	
1		
X		Plain Weave
	X	
X		
	X	

Warp

8/2 cotton: Flax

Total ends: 300

Sett: 20 epi

Width: 15 inches

Weft

Brassard Cotton Bouclé: Black and Yellow

The hooked rug that inspired my version

ACKNOWLEDGMENTS

You know, a book doesn't just happen overnight, and it doesn't get written alone. Yes, my name is on the cover, but this book couldn't have happened without the help of many individuals behind the scenes. Some are actively involved in my life each day, and others, well, I don't even know their names, but they inspired designs in this book. One woman was simply wearing a striking jacket at a concert we attended. She sat five rows in front of us, and I was quick to write down an idea on an old receipt in my wife's purse. I thank Cindy for saving receipts, and I promise never to question why she does that.

Thank you, Stackpole Books and Candi Derr, for believing in me and giving me the two-plus years it took to complete this manuscript. Thank you, Patricia Stevenson, my production editor, for bringing all the pieces together to make such a wonderful book. Thank you, Kathy Eckhaus, for all your wonderful photography and for making my ideas clearer to the readers. You sometimes risk your own health by standing on tall ladders to get just the right shot for a piece. You also volunteered to weave some of the samples for this book. That's a real friend. Your knowledge of weaving and textiles makes you the best with a lens and a loom.

Thanks also go to Vonnie Davis. When we talk about weaving and Vonnie gets excited about a project idea, she always volunteers to weave a sample. Thank you, Vonnie, for your beautiful towel contribution.

To my beautiful wife, Bink (aka Cindy), thank you for wearing so many hats during this journey of clasped weft. You have been my assistant weaver, finisher, cheerleader, proofreader, and rudder to keep me on track. You're honest and truthful and gently let me know when I need to change something to make it clearer and more understandable. You're also very knowledgeable about weaving, and when you see something that's not correct, you tenderly let me know that I messed up. For these reasons, I am eternally grateful to you and look forward to many more journeys with you. You hold my hand and make me laugh each day. I am truly blessed, and I thank you.

Thank you to all the inspiring weavers whom I have never met to thank in person. Some are living, while others laid their shuttles down many, many years ago. Their beautiful works of clasped weft have been inspiring to me. The beauty in their simple weaving and playful designs has sparked my imagination. Thank you all.

Don't Miss These Other Great Books by Tom Knisely!

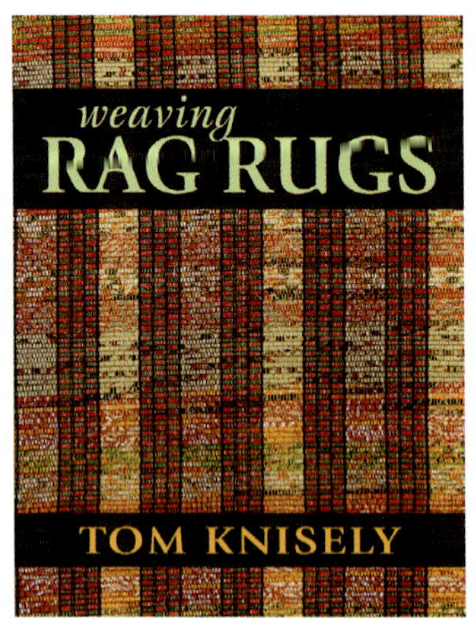

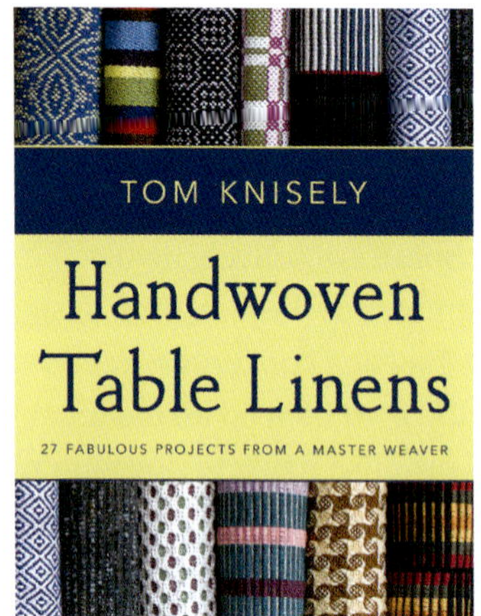

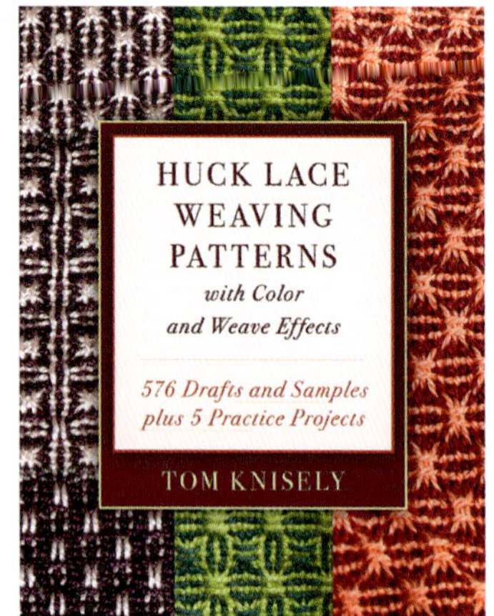

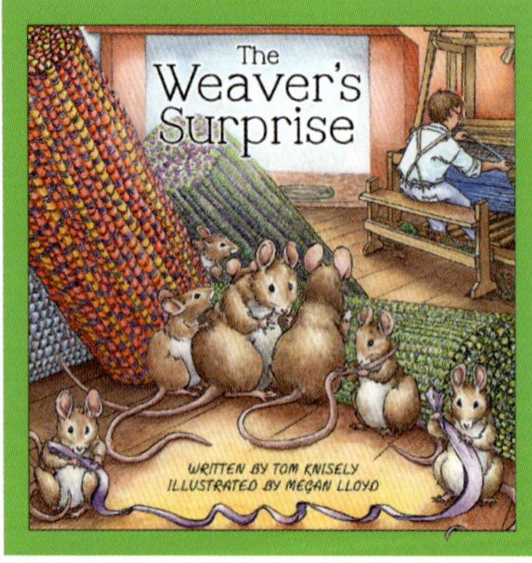

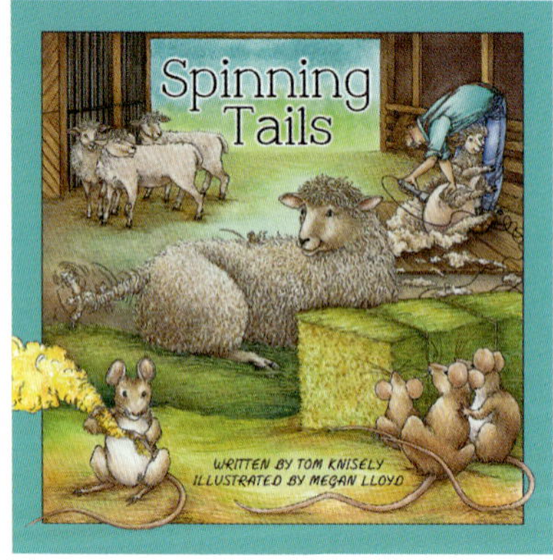

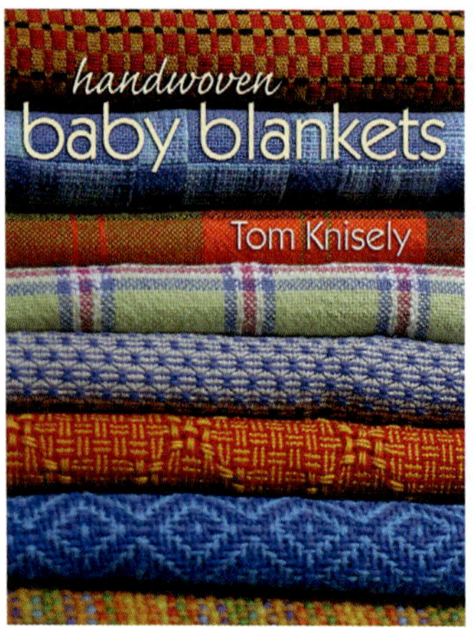

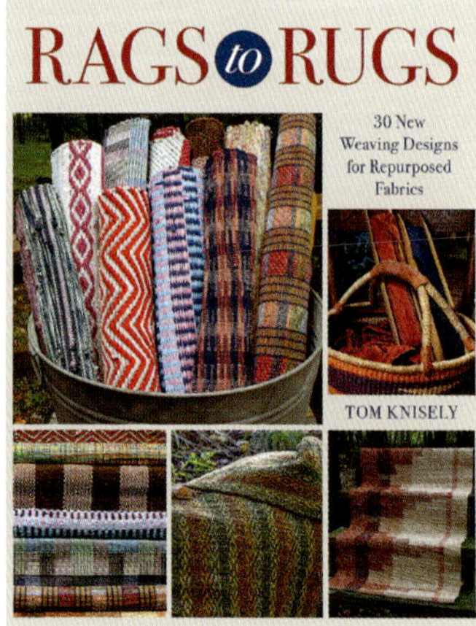